Mihaela Ulieru   Peter Palensky
René Doursat (Eds.)

IT Revolutions

First International ICST Conference
IT Revolutions 2008
Venice, Italy, December 17-19, 2008
Revised Selected Papers

 Springer

Volume Editors

Mihaela Ulieru
The University of New Brunswick
Faculty of Computer Science
Fredericton, New Brunswick, E3B 5A3 Canada
E-mail: ulieru@ucalgary.ca

Peter Palensky
University of Pretoria
Department of Electrical, Electronics
and Computer Engineering
Pretoria 0002, South Africa
E-mail: palensky@ieee.org

René Doursat
Ecole Polytechnique
Institut des Systèmes Complexes
75005 Paris, France
E-mail: rene.doursat@polytechnique.edu

Library of Congress Control Number: 2009932738

CR Subject Classification (1998): J.3, K.4, J.4, H.4, H.5, C.2, I.2.9

ISSN   1867-8211
ISBN-10   3-642-03977-4 Springer Berlin Heidelberg New York
ISBN-13   978-3-642-03977-5 Springer Berlin Heidelberg New York

springer.com

© ICST Institute for Computer Science, Social Informatics and Telecommunications Engineering 2009
Printed in Germany

Typesetting: Camera-ready by author, data conversion by Scientific Publishing Services, Chennai, India
Printed on acid-free paper   SPIN: 12732234   06/3180   5 4 3 2 1 0

Preface

"Mitigating Paradox at the eSociety Tipping Point"

In the first two decades of the past Century, having as driving factor the automobile and its mass production, the command economy has radically changed our lifestyles, enabling the creation of offices, suburbs, fast food restaurants and unified school districts. With the Internet as driving factor, socio-technical and industrial eNetworked ecosystems are about to change our lives again in these two decades of the twenty-first century, and we are just approaching the tipping point. As we have just reached the point where the tremendous changes fueled by concerted efforts in information communication technologies (ICT) research are unraveling the old society this is creating a lot of discomfort, confusion and sometimes opposition from the traditional mainstream. This disconnect is being deepened even more by the rocketing speed of technological ICT advances. As technology is getting ahead of society, the old ways, although still dominant, become more and more dysfunctional and we are experiencing an "age of paradox" as the new ways disrupt the way we used to do things and even the way we used to think about the world. Just like the major inventions that shaped the last century were made by 1920, it is expected that the major inventions that will shape the twenty-first century are going to be made by 2020. Thus, time is quite ripe to ask ourselves, as we witness how the new ways irreversibly challenge the old: Where are the limits of the impossible? And are we ready to touch them with the same daring attitude which fuels our drive to push continuously the technological frontiers? As a premier conference of the ICST, IT Revolutions will account yearly on the progress made in building a critical mass for the radical shift while underlining the major issues and identifying gaps between the traditional and the new ICT-driven ways to provide strategic directions in addressing them—thus ensuring a smooth transition toward an IT-driven future. For this first edition we assembled a brilliant team clustered around major themes concerned with the disconnect between old and new ways reshaping critical aspects of our world at this turning point—in search for innovative solutions that can capitalize on the impact of the tumultuous transformations.

A major characteristic of the IT revolution which the ICST is championing internationally is that technology is far ahead of the yet unused potential for innovations—and this is mainly because the traditional ways are so drastically challenged by the novel ways, and that at the moment the old acts as a barrier rather than an enabler for implementing the technological advances into practice. To mediate this disconnect while embarking responsibly on the momentum that will be turning the world around, in this first year we need to:

- Evaluate where we are on the path to turnover. As this requires coordinated efforts, government S&T representatives and research funding agencies from all continents addressed the major areas of concern in dedicated panels whose deliberations are published as joint recommendations.

- Ensure a smooth transition. Exploring the core elements needed to encourage sustainable change as new systems are introduced and grounded in the practice environment. Motivational factors, incentives to embrace the change as well as alignment with the societal needs were debated in panels and workshops.
- Point to the future. Foresight presentations from most visionary contributors to the major areas helped push the limits of the achievable through keynotes and position papers that underline factors enabling the IT revolution to redesign the world economy and society.

I am certain that you will feel as compelled as we are to become a part of this quest. Thus I invite you to join our concerted efforts and contribute your share to the IT Revolution.

Mihaela Ulieru
Peter Palensky
René Doursat

Organization

Honorary Chair

Imrich Chlamtac   President, Create-Net Research Consortium

General Chair

Mihaela Ulieru   Canada Research Chair in Adaptive Information
                                 Infrastructures for the eSociety Director, Adaptive
                                 Risk Management Laboratory Professor, Faculty of
                                 Computer Science The University of New Brunswick

Technical Program Co-chairs

Rene Doursat   Institut des Systèmes Complexes, Paris, France
Peter Palenski   University of Pretoria, South Africa

Technical Program Committee

Paradox in Convergence

Jack Smith   Defence R&D Canada

Paradox in Health Care

Christian Couturier   Director General – National Research Council
                                     Institute for Information Technology, Canada

Paradox in Schooling: EDU 2.0

Phillip Long   MIT - USA and University of Queensland, Australia

Paradox in eCommerce

Neel Sundaresan   Sr. Director and Head, eBay Research Labs

Paradox in 21st Century Culture: Re-wiring Art, Media and Ethics

Ebon Fisher   Stevens Institute of Technology, USA

Paradox in Approaching Reality

Andres Sousa-Poza   USA

Paradox in Safety and Security

Tim Rosenberg   President and CEO White Wolf Security, USA
Jack Smith   Defense R&D Canada

Paradox in Paradox in Urban Development: The Ubiquitous City

Unho Choi   Executive Director, Korea Logistics Network Corp.,
                          Korea

Paradox in Climate Change

Richard Somerville   USA

Paradox in Approaching Complexity

Yaneer Bar-Yam   President New England Complex Systems Institute,
                          USA
Rene Doursat   Institut des Systèmes Complexes, Paris

Paradox in Social Interaction

Professor David Benyon   Napier University, Scotland, UK

Paradox in Socio-Economic Systems

Mike Hollinshead   Facing-the-Future, Canada; and Walter
                          Derzko, Smart Economy, Toronto, Canada

Paradox in Cyber-Physical Systems

DK Arvind, Director   Research Consortium in Speckled Computing, UK

Paradox in Industrial Automation: Automation 2.0

Jose Luis Martinez Lastra Tampere University of Technology, Finland

Paradox in Nanotechnology

Allan Syms   CEO Ant Nano, UK

Paradox in eNetwork Design

Peter Chapman   Canada

Paradox in Computing

Wolfgang Gentzsch   Director At-Large, Board of Directors, Open Grid
                                 Forum, Germany

Paradox in AI - AI2.0.: The Way to Machine Consciousness

Peter Palensky   University of Pretoria, South Africa

Table of Contents

Position Papers

NRC Papers

IT Complexity Revolution: Intelligent Tools for the Globalised World Development

Andrei Kirilyuk[*] and Mihaela Ulieru

Adaptive Risk Management Laboratory, Faculty of Computer Science
University of New Brunswick, P.O. Box 4400 Fredericton
New Brunswick, E3B 5A3 Canada
Andrei.Kirilyuk@Gmail.com, ulieru@unb.ca

Abstract. Globalised-civilisation interaction intensity grows exponentially, involving all dimensions and regions of planetary environment. The resulting dynamics of critically high, exploding complexity urgently needs consistent understanding and efficient management. The new, provably universal concept of unreduced dynamic complexity of real interaction processes described here provides the former and can be used as a basis for the latter, in the form of "complexity revolution" in information systems controlling such "critically globalised" civilisation dynamics. We outline the relevant dynamic complexity properties and the ensuing principles of anticipated complexity transition in information and communication systems. We then emphasize key applications of unreduced complexity concept and complexity-driven IT to various aspects of post-industrial civilisation dynamics, including intelligent communication, context-aware information and control systems, reliable genetics, integral medicine, emergent engineering, efficient risk management at the new level of socio-economic development and resulting realistic sustainability.

Keywords: Complexity, chaos, self-organisation, fractal, adaptability, dynamic multivaluedness, emergent engineering, adaptive risk management.

1 Introduction

Exponentially growing power of various interaction processes within increasingly globalised civilisation has surpassed today the characteristic *generalised globalisation* threshold, after which everything becomes related to everything else and the number of unconditionally negligible interaction links tends to zero. A major characteristic of such "truly globalised", strong-interaction system structure and dynamics is that they cannot be a-priori defined - but rather emerge from the "bottom-up" interactions between component systems (and people). This clashes with the traditional systems engineering that approaches the design of hard- and software "with the end in mind", namely by a-priory defining the system and its performance requirements following a "top-down", linear thinking. Given the exponentially huge, always growing number of possibilities, such approach cannot be efficient above globalisation threshold, for

---

[*] On leave from Institute of Metal Physics, 36 Vernadsky Avenue, 03142 Kiev, Ukraine.

M. Ulieru, P. Palensky, and R. Doursat (Eds.): IT Revolutions 2008, LNICST 11, pp. 1–13, 2009.

any linear, basically sequential computing power. It is interesting to note that there is a similar contradiction in fundamental science between predetermined configuration of its "exact-solution", or "perturbative", constructions and much more variable, emergent diversity of real-system structure it is supposed to describe.

In order to cope with this fundamental difficulty and realise the new, *emergent engineering* approach [1], one should first understand the *unreduced interaction dynamics* within a real system, including both controlled natural/human systems and controlling information systems supposed to correctly reproduce main features of controlled system behaviour. Special attention should be given here to processes of new, a-priori unknown structure formation, so that one could eventually replace explicit system design with definition of final purposes and general criteria (that should also be rigorously and universally specified).

The unreduced, real-system interaction description with the necessary properties has been proposed recently for arbitrary kind of system, leading to the *universal concept of dynamic complexity* that was then confirmed by application to various particular systems, from fundamental physics to autonomic communication and information systems [2-6]. We shall outline below major features of those results having strategic importance for IT Revolutions programme (Sect. 2) and showing that to a large degree it can be properly specified in terms of *IT complexity revolution* implying indeed a qualitatively big transition from zero-complexity to high-complexity, intelligent and autonomic ICT structures (Sect. 3). While this *complexity transition* in artificial structures remains a real *challenge* we face today, the properties we should obtain and their origin in natural structures are clearly specified indicating ways to problem solution and providing various intuitive expectations about them with a rigorous basis and quantitative criteria. We conclude by demonstrating, in Sect. 4, inevitability of ICT complexity revolution for key applications, such as reliable genetics, integral medicine and new, complex-dynamic eNetwork of creatively monitored, decentralized and sustainable production processes [7,8].

2 Unreduced Interaction Dynamics: Causal Randomness, Emerging Structure and Unified Complexity Definition

Arbitrary interaction configuration, including that of information-exchanging entities within an autonomic eNetworked system, can be *universally* represented by a Hamiltonian-form equation we call here *existence equation* because it simply fixes the fact of many-body interaction within a system, without any special assumption [2-6]:

$$\left\{ \sum_{k=0}^{N} \left[h_k(q_k) + \sum_{l>k}^{N} V_{kl}(q_k, q_l) \right] \right\} \Psi(Q) = E\Psi(Q), \tag{1}$$

where $h_k(q_k)$ is the "generalised Hamiltonian" for the k-th system component, q_k are its degrees of freedom, $V_{kl}(q_k, q_l)$ is the (arbitrary) interaction potential between the k-th and l-th components, $\Psi(Q)$ is the system state-function, $Q \equiv \{q_0, q_1, ..., q_N\}$, E is the generalised Hamiltonian eigenvalue, and summations include all (N) system components. Note that explicitly time-dependent system configurations (with time-dependent interaction potentials) are described by the same equation, up to formal notation change, but here we want to concentrate on time-independent initial configuration, in order to emphasize *emergent* time (and structure) effects.

A usual, perturbative approach to system behaviour analysis involves essential reduction of generally unsolvable (nonintegrable) Eq. (1) to a formally solvable one, but missing many real-system interaction links, such as

$$[h_0(\xi) + \tilde{V}_n(\xi)]\psi_n(\xi) = \eta_n\psi_n(\xi),$$

(2)

where $\xi = q_0$ is one of system degrees of freedom, $\psi_n(\xi)$ is a state-function component, and the "mean-field" potential

$$|V_{nn}(\xi)| \ll |\tilde{V}_n(\xi)| \ll |\sum_{n'} V_{nn'}(\xi)|.$$

(3)

It is by choosing a particular configuration of this extremely simplified mean-field potential that one artificially imposes the "expected", predetermined interaction result, any genuine emergence (structure-formation, or self-organisation) effects being practically cut off, irrespective of further "stability analysis" or "peculiar" trajectory behaviour.

However, there is a way to efficiently analyse such "intractable" (practically all real!) interaction problems without their unjustified, creation-killing simplification. It can be done in the (properly extended) framework of so-called optical, or effective, potential method [2-6], where one obtains an equation formally similar to approximate Eq. (2) but being, contrary to the latter, equivalent to the unreduced problem formulation, Eq. (1), in its full complexity:

$$h_0(\xi)\psi_0(\xi) + V_{\text{eff}}(\xi;\eta)\psi_0(\xi) = \eta\psi_0(\xi),$$

(4)

where η is the problem eigenvalue to be found, while the *effective potential (EP)* $V_{\text{eff}}(\xi;\eta)$ *depends* now, in a very complicated and *highly nonlinear* way, on this eigenvalue and eigenfunction $\psi_0(\xi)$ to be found (details can be found in Refs. [2-6] and other papers cited therein). This EP structure actually contains, now in a more convenient (though always formally "nonintegrable"!) form, the full complexity of all system interactions, their emerging links, possible "round-about ways", etc. It is not difficult to show that this nonlinear, "additional" EP dependence on the eigen-solutions to be found leads to real existence of *multiple*, locally *complete* and therefore *mutually incompatible* problem solutions there where one would expect only one its solution within any perturbative or exact-solution approach of Eqs. (2)-(3). Therefore these mutually incompatible but *equally* real solutions called system *realisations* (they describe its various possible, now *explicitly emerging* configurations) are forced, by the driving interaction itself, to *permanently replace one another*, in a *dynamically (causally) random* order thus defined.

The phenomenon of *dynamic multivaluedness*, or *redundance*, thus *rigorously derived* by the *unreduced* interaction analysis changes dramatically the richness and quality of system behaviour (it should be distinguished from usual "multistability" or "strange attractors" obtained within the reduced, dynamically single-valued analysis equivalent to a mean-field approximation of Eqs. (2)-(3)). First of all, it provides the *universal and strictly intrinsic (dynamic) origin (and the very meaning) of randomness* in *any* real interaction process, which does *not* depend on (formally postulated) time or any other external factors (like "initial conditions" for "diverging trajectories"). It is confirmed by the related *dynamic* definition of *a-priori* event probability, which is not related to any abstract, postulated "event space" but directly follows from the above random change of *dynamically(objectively) equal* realisations:

$$\alpha_r = \frac{N_r}{N_{\Re}} \left(N_r = 1, \ldots, N_{\Re}; \sum_r N_r = N_{\Re} \right), \quad \sum_r \alpha_r = 1 , \qquad (5)$$

where α_r is the probability of *dynamic emergence* of r-th actually observed realisation containing N_r elementary realisations ($N_r = 1$ for each of these), while $N_{\Re}$ is the (total) *number of system realisations*. This emergent randomness and probability phenomenon shows, in particular, that any hope for at least theoretically possible regularity of "pure" (properly isolated) and "thoroughly controlled" interaction processes (also in ICT systems) is vain: the origin of randomness is *within* even the formally totally regular, "deterministic" interaction itself (even for zero-uncertainly "initial conditions"!), while any additional "control" attempt is but a new interaction configuration subject to the same, intrinsic randomness.

Directly related to this result is the *universal, dynamic event definition*, which is nothing but each subsequent *realisation (or their dense group) emergence/change* in the process of their permanent "competition" (due to dynamic multivaluedness) within initially "homogeneous" and *time-independent* interaction process. This is the *rigorously specified* process of structure formation, or emergence, or self-organisation, that *cannot* be specified in principle within any usual, *dynamically single-valued* analysis (because the single system realisation just remains identical to itself, without any intrinsic change). It is confirmed by the inevitably related, rigorously derived definition of physically real, emergent, *unceasingly and irreversibly flowing time*. Namely, time acquires an elementary increment as a result of strongly inhomogeneous *realisation change event*, which occurs in our *initially totally timeless* system due to the same dynamic multivaluedness phenomenon. Each real system jump (dynamic reconstruction) from one of its *multiple* realisations (configurations) to another, *randomly chosen* one gives rise to the related *quantum of space* (for a given interaction level), or characteristic (minimum) size, Δx, directly determined by *eigenvalue spacing* of the unreduced EP problem, Eq. (4), $\Delta x = \Delta_r \eta_i^r$ [2-6], after which the elementary time increment Δt is obtained as $\Delta t = \Delta x / v$, where v is the speed of (homogeneous) signal propagation in physically real system "environment" (one of its initial degrees of freedom in Eq. (1)). Unstoppable *time flow* is due to unceasing *realisation change*, while its intrinsic *irreversibility* is due to the *causally random* order of their appearance (time is thus inseparable from dynamic randomness of time-making event emergence). It is evident that this rigorously specified, physically "produced", quantised space and irreversibly flowing time have a *hierarchical structure*, reappearing (emerging) at each new interaction level as relevant-scale entities dynamically "constructed" from effectively homogeneous degrees of freedom of previous, lower level(s).

And finally, one obtains as a unifying result the provably consistent and totally universal definition of *dynamic complexity, C* (and closely related *chaoticity*):

$$C = C(N_{\Re}), \; dC/dN_{\Re} > 0, \; C(1) = 0, \tag{6}$$

where $N_{\Re}$ is the system realisation number (determined eventually by the number of its interacting components, N [2-6]) and (dynamic) complexity is universally measured by any *growing* function of realisation number, $C(N_{\Re})$, or its rate of change, equal to zero for (actually unrealistic) case of $N_{\Re} = 1$ (usually $N_{\Re} \geq N \gg 1$). Note that it is actually the last, totally unrealistic case of zero-complexity, dynamically single-valued interaction result (problem solution) that is invariably considered within usual "models" (including those used in "complexity science", "chaos theory", etc.) that can only provide *effectively zero-dimensional*, point-like "projections" of dynamically multivalued, complex-dynamic behaviour of real systems. Those projections can certainly bear various "signatures" of underlying real-system complexity, but in its strongly and *unpredictably* reduced version. Note, in particular, that $N_{\Re}$ in Eq. (6) stands for the number of *explicitly obtained* (as a result of unreduced problem solution, Eqs. (1), (4)), mutually incompatible and changing system realisations, rather than the number of arbitrary observed and "countable" entities. Suitable complexity measures include $C(N_{\Re}) = C_0 \ln(N_{\Re})$, $C(N_{\Re}) = N_{\Re} - 1$, or $\partial N_{\Re}/\partial t$. As dynamic multivaluedness underlies both genuine dynamic complexity and causal randomness (see Eq. (5)), we can also define complex behaviour as *chaotic* (dynamically random) one, (dynamical) *chaos* being consistently and universally specified now as (always) causally random process of system *realisation change*, which is the only possible way of any real system or interaction process *existence*. It shows once again that attempts to establish total regularity in any real, complex system cannot be successful in principle, especially in higher-complexity (large and autonomic) systems of our main interest (see also below).

According to intrinsically probabilistic origin of any emerging system configuration, any measured quantity represented by the generalised system density $\rho(Q)$, is obtained as a *causally probabilistic* sum of respective quantity values for individual realisations, $\rho_r(Q)$:

$$\rho(Q) = \sum_{r=1}^{N_{\Re}} {}^{\oplus} \rho_r(Q), \tag{7}$$

where detailed expressions for $\rho_r(Q)$ can be obtained within the unreduced EP method using Eq. (4) [2-6] and the sign $\oplus$ designates the special, dynamically probabilistic meaning of the sum. It implies that the observed quantity $\rho(Q)$ permanently, randomly changes, together with system realisation, between its $N_{\Re}$ possible values $\{\rho_r(Q)\}$ appearing with their *dynamically determined probabilities* $\{\alpha_r\}$ of Eq. (5). The dynamic origin of probability thus obtained means also that, contrary to usual situation, the result of Eqs. (5), (7) remains valid irrespective of the number of observed events (observation time) and, in particular, it is valid for every single realisation (event) emergence (and even *before* any event emergence!).

However, the complete problem solution has even more complicated structure than that of the causally probabilistic sum of Eq. (7) representing only the first level of system dynamics splitting into many incompatible, changing realisations. Each of these realisations generally shows similar internal splitting into second-level, also incompatible and probabilistically changing realisations (under the influence of the same system interaction) and so on, where the total number of such realisation levels can be very large even though not all of them can be practically resolved as such. This general, most complete problem solution (expressed by a *multi-level* probabilistic sum in Eq. (7)) has thus the structure of *dynamically probabilistic fractal* [2-4], whose multi-level hierarchy of ever finer elements is much richer than that of usual fractals because it includes permanent *probabilistic realisation change* at each structure level, providing it with the property of *efficient dynamic adaptability*, or *intelligent (reasonable) behaviour*, observed in *living* organisms.

Useful *power* P_{real} of that probabilistic fractal dynamics underlying such "magic" properties as high adaptability, autonomy, creativity, intelligence and sustainable development (highly desired for the new ICT and eSocial systems!) is determined by the total number of (fractal) realisations $N_{\Re}$ (or complexity C) that can be estimated as the *number of system link combinations* [3-6]:

$$P_{real} \propto N_{\Re} \sim L! \approx \sqrt{2\pi L}\left(\frac{L}{e}\right)^L \sim L^L \propto C, \tag{8}$$

where the number of links L is already very large (it can be much greater than the number of system components N: for human brain, $N > 10^{10}$, $L > 10^{14} \gg N$). Useful power of corresponding systems with traditionally limited (regular, sequential, linear) operation can at best grow only as $L^\beta \ll P_{real}$ ($\beta \sim 1$). It is this *exponentially huge* efficiency advantage that explains the above "magic" *qualities* of high-complexity (very large L) natural systems (*life, intelligence, consciousness, sustainability*), which can be successfully reproduced and efficiently controlled in man-made, artificial environment only if one "liberates" the involved information (and human!) system dynamics to follow a creative, free-interaction regime.

Every structure-formation, emergent system dynamics resulting from unreduced, complex-dynamic interaction development can also be called *self-organisation*. Although we have shown above that any really emerging, self-organised structure inevitably contains a great deal of dynamic randomness, the latter is usually *confined* to a more or less distinct shape determining the observed system configuration. How can one define then the actual "proportions" of, and the border between, those omnipresent but opposite properties of randomness and order? The detailed analysis of EP method equations (see Eq. (4)) shows [2-6,9] that the onset of strongly irregular regime of *uniform chaos* occurs under the condition of *resonance* between major component processes, such as the internal dynamics of each system component and characteristic interaction transmission between components:

$$\kappa = \frac{\omega_\xi}{\omega_q} \cong 1, \tag{9}$$

where κ is the introduced *chaoticity* parameter, while ω_ξ and ω_q are frequencies (or energy-level separations) for the inter-component and intra-component motions, respectively. At $\kappa \ll 1$ (far from resonance) one has a multivalued self-organised or confined-chaos regime (internally chaotic but quasi-regular externally), which becomes the less and less regular as κ grows from 0 to 1, until at $\kappa \approx 1$ (resonance condition) the global or uniform chaos sets in, followed by another self-organised regime with an "inverse" system configuration at $\kappa \gg 1$. All multi-level hierarchy of any real system dynamics can be described within this universal classification, where additional complication comes from the fact that there are always many higher-order resonances in the system (describing always present by maybe spatially limited chaoticity). In natural system dynamics the regimes of uniform chaos and self-organisation tend to "reasonably" alternate and coexist, so that the former plays the role of efficient "search of the best development way", while the latter ensures more distinct structure creation as such, both providing detailed realisation of the above high interaction power, Eq. (8), to be reproduced in the next generation of intelligent eNetworks, where the quantitative criterion of Eq. (9) can be useful as a universal guiding line (uniquely related to the above unreduced complexity analysis).

The qualitatively strong, irreversible change and creativity inherent to the unreduced, multivalued system dynamics imply certain *direction* and *purpose* of real interaction processes. These can be specified [2-6, 10] as *unceasing* transformation, in *any* interaction process and system dynamics, of a *latent* complexity form of *dynamic information*, I (generalising the notion of "potential energy"), into its *explicit* form of *dynamic entropy*, S (generalising the notion of "kinetic energy"), always occurring so that the *total complexity*, C, given by the sum of the two complexity forms, decreasing dynamic information and increasing dynamic entropy, $C = I + S$, remains *unchanged* (it is given by the initial interaction configuration and any its external modification):

$$\Delta C = 0, \quad \Delta S = -\Delta I > 0. \qquad (10)$$

Dynamic information describes system potential for new structure/quality creation, while dynamic entropy describes the unreduced dynamic complexity of already created structure. The *universal symmetry of complexity* (law of its conservation and transformation), Eq. (10), provides thus another useful guideline for emergent engineering by rigorously specifying the *universal purpose of natural interaction development*. One can show that all major laws and dynamic equations known from fundamental physics can be reduced to particular cases of this universal symmetry of complexity [2,3,10], which is further evidence in its favour as a guiding rule.

3 Complexity Transition: Creative ICT Systems and Emergent Engineering

Rigorous description of unreduced interaction dynamics from the previous section including provably universal concept of complexity can be considered as a necessary basis for the new, *exact* science of intelligent and creative ICT tools and their efficient application to management and development of complex real-world dynamics. That kind of theory provides a *rigorously specified* extension of respective empirical results and intuitive expectations about the next stage of ICT development confirming its now *provably revolutionary* character and specifying its *objectively efficient* content. One should also take into account that such complex information systems are supposed to be used for efficient control of real, high-complexity systems, implying consistent, realistic and universal understanding of controlled system dynamics (otherwise missing) as indispensable condition for their sustainable management.

Major features of unreduced complex dynamics relevant for the IT complexity revolution were outlined in Sect. 2: huge power of unreduced interaction process, its inevitable and purely dynamic randomness, universal classification of major regimes of truly complex dynamics, and the universal symmetry of complexity as the unified law and purpose, as well as guiding line for ICT system design. Now we can further specify these results in terms of several major *principles of complex ICT system operation and design* [6] realising also the ideas of emergent engineering [1].

We start with the *complexity correspondence principle* implying efficient interaction only between systems of *comparable* (unreduced) complexity. Being a direct consequence of complexity conservation law, Eq. (10), it limits the scope of efficient system design to cases that do not contradict that fundamental law and therefore can be realised at least in principle (similar to energy conservation law use in usual, thermo-mechanical machine construction). A major manifestation of the complexity correspondence principle is the "complexity enslavement rule" stating that a higher-complexity system can efficiently control (or enslave) a lower-complexity one, but never the other way around. Therefore, there is no sense to try to obtain efficient (autonomic) control over a complex system using only lower-complexity tools. It can be considered as rigorous substantiation of the importance of the whole IT complexity revolution concept we describe here: in its absence, traditional, zero-complexity systems monitoring complex real-world phenomena can be used only in a strongly non-autonomic way implying essential input from human complexity levels (intelligence), but in today's increasingly "globalised", strong-interaction world efficient application of such man-dominated, "slow" and "subjective" control becomes ever less efficient if not catastrophic.

According to the same principle, interacting similar-complexity systems may easily give rise to strongly chaotic (dynamically multivalued) behaviour that can be both harmful (in situations where one is looking for stability of a basically established configuration) and useful (in situations where essentially new and best possible ways of complexity development should be found). And finally, when one tries to use a very high-complexity system for control over a much lower-complexity one, this control should certainly be possible, but may be practically inefficient for another reason: the high-complexity controller will effectively "replace" (or even suppress) the much lower-complexity but useful process under its "surveillance". This would imply, in particular, that IT complexity revolution should better start and proceed from software/context to hardware/traffic level, rather than in the opposite direction.

These rules of efficient complexity control can be extended to the second major principle of complex ICT management, the *complex-dynamic control principle*. Based on the same universal symmetry of complexity, but now rather its complexity development aspects (see the end of Sect. 2), it states that contrary to ideas of traditional, "fixed" control, the extended, complex-dynamic control implies suitable *complexity development* (i.e. *partially unpredictable change*) as a major condition for efficient system monitoring. It means that, as proven by our unreduced interaction analysis (Sect. 2), controlled dynamics can *not* - and should not - be totally, or even mainly, regular. In other words, suitable degrees of acceptable chaotic change, or even essential structure development, should be provided for truly efficient, failure-proof eControl systems (involving both their ICT and human elements.)

The purpose of such extended, complex-dynamical control becomes thus practically indistinguishable from the general direction of *optimal interaction complexity development* (from dynamic information to dynamic entropy, Sect. 2), as it should be, especially in a "generally globalised" system, taking into account its inseparable interaction structure including both controlling and controlled elements. It shows that properly *creative* monitoring is actually much more *reliable* than traditional, restrictive control, implying once again the necessary essential advance towards complex-dynamic, truly intelligent IT control systems. Combined with the first principle of complexity correspondence and enslavement, this universal guiding line and criterion of complex-dynamic control leads to the ultimate, now uniquely realisable purpose of *sustainable control* providing (unlimited) *global development stability* by way of omnipresent *local creativity*.

Finally, the *unreduced (free) interaction principle* refers to exponentially huge power of unreduced interaction processes, as opposed to much lower, power-law efficiency of traditional, linear (sequential) operation schemes in existing ICT systems (see Eq. (8) in the previous section). The above complexity correspondence principle confirms the evident fact that efficient management of that huge real-interaction power would need equally high, complex-dynamic interaction power of IT systems applied and ever more densely inserted in the tissue of real-world complexity and intelligence. However, such huge power progress needs equally big transition from detailed step-by-step programming in usual ICT approach to monitoring of only general development direction of complexity-entropy growth (by properly specifying complexity-information input). Intrinsic chaoticity should become a normal, useful operation regime. Correspondingly, the huge power of free-interaction dynamics can be realised in the form of dynamically probabilistic fractal (Sect. 2), with specially allocated possibilities of its development. As mentioned above, such natural complexity development tends to occur as irregular, constructive alternation of global chaos and multivalued self-organisation regimes of complex dynamics, the former realising efficient search of optimal development ways and the latter providing a more ordered structure creation as such.

The unified, rigorously substantiated content of these three major principles of complex ICT system operation and design reveals the forthcoming ICT *complexity transition* as the first stage of IT complexity revolution and the beginning of *useful* complexity and chaos role in information-processing systems. This qualitatively big transition will show up as essential power growth and appearance of new features usually attributed to living and intelligent systems. Whereas technical realisation of complexity transition remains an exciting challenge, the objective necessity to meet it follows e.g. from the universal criterion of chaos as being due to system resonances (see Eq. (9)). While a sufficiently low-intensity eNetwork can try to maintain its basic regularity by avoiding major resonances, it becomes impossible for higher network interaction density/intensity due to inevitable overlap of emerging new resonances. Therefore even a limited task of preserving traditional system regularity acquires a nontrivial character and needs application of unreduced complexity description. While everything shows that today we are already quite near that complexity-transition threshold, the unreduced interaction complexity opens also much brighter perspectives of new, generally unlimited network possibilities after successful complexity transition, providing additional motivation to the whole problem study. As strong interaction within the global ICT-human-social system is inevitable in any case and has already been realised in large parts of the world, its often negative aspects of "future shock" [11] (where dense but regular IT systems tend to "impose" their "pathological" linearity to intrinsically complex human thinking) can be replaced by the opposite positive effects only as a result of IT complexity revolution (where human intelligence endows IT systems with a part of its natural complexity).

One relatively easy way to approach the desired complexity transition starting from existing network structure is to attempt a transition from their still hardware- and location-based realisation to intelligent-software- and *knowledge-based* realisation. Such truly knowledge-based networks can be conceived as autonomic systems of interacting and permanently *changing* knowledge (any semantic) structures able to usefully evolve *without* direct intervention of human user (that will instead create and modify general rules and particular preferences of this knowledge interaction process). Efficiency demands for such knowledge-based ICT development will naturally enforce the advent of complex-dynamic operation modes, simply due to complex-dynamic structure of unreduced knowledge content. This major example shows how even a "usual" quality-of-service demand involves complexity transition in ICT system operation and design. This is also the next, equally natural step of emergent engineering [1] realisation in its *autonomic engineering* version, where the omnipresent, real-time (and knowledge-based!) system development constitutes an integral part of its complex, holistic dynamics within any particular application.

4   IT Revolutions as Complexity Revolution in Science and Society

The IT complexity revolution substantiated and specified in previous sections as the core of modern IT revolution represents a natural result of ever more interactively and intensely used information and communication systems approaching now the critical point of complexity transition (Sect. 3). This rapidly advancing and still poorly recognised process acquires yet greater importance and support if one takes into account equally big complexity transitions emerging in all key fields of science, technology and global civilisation dynamics constituting together the forthcoming *complexity revolution* in human civilisation development [5]. It becomes evident that all these essentially complex-dynamical applications need complex-dynamic IT tools for their real progress (due to the rigorously substantiated complexity correspondence principle, Sect. 3), while successful development of those tools can only result from complexity science and application progress. It would be not out of place to briefly outline here the key complexity applications asking for IT complexity revolution.

While today's rapid progress of *genetics* and related bio-medical applications is widely acknowledged, emerging serious problems around the unreduced dynamic complexity of bio-chemical systems involved only start appearing in public science discussions (see e.g. [12]). The universal science of complexity clearly specifies the irreducible origin of those problems [3] in terms of exponentially huge power of real interaction processes, with characteristic values of the number of essential interaction links $L \sim 10^{12} - 10^{14}$ in Eq. (8), making any usual (sequential/regular) computing power negligible with respect to practical infinity implied by the estimate of Eq. (8). It follows that only essential progress towards equally rich, truly complex dynamics of information systems used for the causally complete understanding of living organism dynamics can solve this kind of problem and in particular form a solid basis for the truly *reliable genetics*.

Very close to genetics is the extremely popular group of *nano-bio applications*, where the fact of strongly chaotic (multivalued) nano-system dynamics [13] remains practically unknown, despite the evident similarity with the above bio-chemical problems. Here too, the dominating incorrect reduction of nano-system dynamics to regular models can be as harmful as the unreduced complexity analysis can be advantageous, asking for complex-dynamic IT systems as major study tools.

At a higher complexity level of the whole living organism dealt with in *medicine*, it becomes evident that any its correct understanding (absolutely necessary for sustainable progress of extended life quality) should involve suitably complex information system dynamics able to provide a unified complex-dynamical "map" of each individual organism dynamics and development. This is the idea of *integral medicine* [2,3] based on inseparable combination of unreduced complexity science and complex-dynamic IT systems applied. Comparing these natural perspectives and challenges with the reductive and separating approach of usual, mechanistic medicine, it is easy to see the necessity of *bio-medical complexity revolution*.

Sustainable development issues represent a further natural extension of the same ideas to ever higher complexity levels of planetary civilisation dynamics involving nevertheless a well-specified conclusion about the necessary unified transition to a superior level of the entire civilisation complexity [5]. This "embracing", *global complexity revolution* can only be avoided in the case of alternative possibility of explicitly degrading, complexity-destroying development branch (that may have already become dangerously real). As knowledge-intense civilisation structure includes, already today, ICT systems as its essential part, the forthcoming global complexity revolution and the next development stage cannot avoid essential use of properly upgraded, complex-dynamical information systems.

One can speak here about *complex eNetworks efficiently controlling decentralized, emergent production and creative consumption systems* [7,8] (in the new sense of "developing" complex-dynamical control of Sect. 3) and effectively replacing today's inefficient financial systems that cannot cope with the real development complexity. Note, by the way, that efficient management of huge volumes of (highly interactive) scientific and technological, all innovation-related information alone would certainly need urgent introduction of knowledge-based, context-aware and thus explicitly complex-dynamic ICT tools.

We shall not discuss here much lower complexity levels of *physical systems*, while noting, however, that even at those *relatively* low complexity levels one is definitely pushed towards the unreduced analysis in terms of real system complexity (instead of conventional zero-complexity "models"), without which "unsolvable" problems and related development impasses accumulate catastrophically in various fields, from particle physics and cosmology to high-temperature superconductivity and nuclear fusion [14], leading to the desperate "end of science".

One should also emphasize a *general-scientific* but practically important meaning of consistent understanding of complex ICT system dynamics. Whereas fields like binary algebra and logics formed a basis for information technology development in a previous epoch, today one deals rather with the problem of (arbitrary) interaction within a real system of many "information bodies" whose causally complete solution just leads to the universal concept of dynamic complexity (Sect. 2). The resulting *complex-dynamic information science* [6] will therefore constitute a rigorous basis for the next stage of fundamental and applied computer science development realising essential progress of scientific knowledge. Complex-dynamic, real-knowledge-based, semantically full and therefore inevitably "uncertain" information entities will replace standard, semantically trivial "bits" from the previous, hardware-oriented level of information science, which is exactly what is needed from the point of view of modern application demands [15-18]. This crucial progress involves a qualitatively deep change in dominating conceptual attitudes and empirical methods as discussed above (Sect. 3).

In conclusion, we have demonstrated that the revolutionary situation in IT development clearly felt today can be consistently and constructively specified as IT complexity revolution implying creation and (practically unlimited) development of qualitatively new ICT systems and applications. Namely, we (1) rigorously proved the inevitability of ICT complexity emergence (Sect. 2), (2) derived major properties of interest of real-system complexity within its universally valid concept (Sect. 2), (3) obtained three major principles of complex-dynamic and intelligent ICT development and related complexity transition (Sect. 3), and (4) demonstrated the application-related need for complex ICT system development, including the global complexity revolution leading to intrinsically sustainable level of social and economic civilisation dynamics (Sect. 4). It is therefore difficult to see another, equally promising and problem-solving alternative to this long-term way of ICT development, with its revolutionary start beginning right now and going to be a major contribution of today's generation to the planetary civilisation development.

References

1. Doursat, R., Ulieru, M.: Emergent Engineering for the Management of Complex Situations. In: Proceedings of Autonomics 2008, Second ACM International Conference on Autonomic Computing and Communication Systems, Turin, Italy, September 23-25 (2008), http://www.cs.unb.ca/~ulieru/Publications/ Ulieru-Autonomics-Final.pdf

2. Kirilyuk, A.P.: Universal Concept of Complexity by the Dynamic Redundance Paradigm: Causal Randomness, Complete Wave Mechanics, and the Ultimate Unification of Knowledge. Naukova Dumka, Kyiv (1997), ArXiv:Physics/9806002, http://books.google.com/books?id=V1cmKSRM3EIC

3. Kirilyuk, A.P.: Complex-Dynamical Extension of the Fractal Paradigm and Its Applications in Life Sciences. In: Losa, G.A., Merlini, D., Nonnenmacher, T.F., Weibel, E.R. (eds.) Fractals in Biology and Medicine, Vol. IV, pp. 233–244. Birkhäuser, Basel (2005), ArXiv:Physics/0502133

4. Kirilyuk, A.P.: Consistent Cosmology, Dynamic Relativity and Causal Quantum Mechanics as Unified Manifestations of the Symmetry of Complexity. Report presented at the Sixth International Conference Symmetry in Nonlinear Mathematical Physics (Kiev, June 20-26, 2005), ArXiv:Physics/0601140

5. Kirilyuk, A.P.: Towards Sustainable Future by Transition to the Next Level Civilisation. In: Burdyuzha, V. (ed.) The Future of Life and the Future of Our Civilisation, pp. 411–435. Springer, Dordrecht (2006), ArXiv:Physics/0509234

6. Kirilyuk, A.P.: Unreduced Dynamic Complexity: Towards the Unified Science of Intelligent Communication Networks and Software. In: Gaïti, D. (ed.) Network Control and Engineering for QoS, Security, and Mobility, IV. IFIP, Vol. 229, pp. 1–20. Springer, Boston (2007), ArXiv:Physics/0603132

7. Ulieru, M., Verdon, J.: IT Revolutions in the Industry: From the Command Economy to the eNetworked Industrial Ecosystem. In: Proceedings of the 1st International Workshop on Industrial Ecosystems, IEEE International Conference on Industrial Informatics, Daejoen, Korea, July 13-17 (2008),
http://www.cs.unb.ca/~ulieru/Publications/
Verdon-paper-formatted.pdf

8. Ulieru, M.: Evolving the DNA blueprint of eNetwork middleware to Control Resilient and Efficient Cyber-Physical Ecosystems. Invited paper at BIONETICS 2007 - 2nd International Conference on Bio-Inspired Models of Network, Information, and Computing Systems, Budapest, Hungary, December 10-14 (2007), http://www.cs.unb.ca/~ulieru/Publications/Ulieru-Bionetics2007.pdf

9. Kirilyuk, A.P.: Dynamically Multivalued Self-Organisation and Probabilistic Structure Formation Processes. Solid State Phenomena 97-98, 21–26 (2004), ArXiv:Physics/0405063

10. Kirilyuk, A.P.: Universal Symmetry of Complexity and Its Manifestations at Different Levels of World Dynamics. In: Proceedings of Institute of Mathematics of NAS of Ukraine, vol. 50, pp. 821–828 (2004), ArXiv:Physics/0404006

11. Toffler, A.: Future shock. Bantam, New York (1984)

12. Gannon, F.: Too complex to comprehend? EMBO Reports 8, 205 (2007)

13. Kirilyuk, A.P.: Complex Dynamics of Real Nanosystems: Fundamental Paradigm for Nanoscience and Nanotechnology. Nanosystems, Nanomaterials, Nanotechnologies 2, 1085–1090 (2004), ArXiv:Physics/0412097

14. Kirilyuk, A.P.: The Last Scientific Revolution. In: López Corredoira, M., Castro Perelman, C. (eds.) Against the Tide: A Critical Review by Scientists of How Physics & Astronomy Get Done, pp. 179–217. Universal Publishers, Boca Raton (2008), ArXiv:0705.4562

15. European Commission R&D Framework Programme, Information Society Technologies, Future and Emerging Technologies, Situated and Autonomic Communications,
http://cordis.europa.eu/ist/fet/comms.htm,
http://cordis.europa.eu/ist/fet/areas.htm

16. Bullock, S., Cliff, D.: Complexity and Emergent Behaviour in ICT Systems. Foresight Intelligent Infrastructure Systems Project. UK Office of Science and Technology (2004),
http://www.foresight.gov.uk/OurWork/CompletedProjects/IIS/
Docs/ComplexityandEmergentBehaviour.asp

17. Di Marzo Serugendo, G., Karageorgos, A., Rana, O.F., Zambonelli, F. (eds.) ESOA 2003. LNCS, vol. 2977. Springer, Berlin (2004)

18. Smirnov, M. (ed.): WAC 2004. LNCS, vol. 3457. Springer, Heidelberg (2005)

"Low Power Wireless Technologies: An Approach to Medical Applications"

Francisco J. Bellido O.[1], Miguel González R.[1], Antonio Moreno M.[1], and José Luis de la Cruz F.[2].

[1] Dept. of Electronics
[2] Dept. of Applied Physics, University of Cordoba, Spain
{fjbellido,el1gorem,amoreno,fa1crfej}@uco.es

Abstract. Wireless communication supposed a great both -quantitative and qualitative, jump in the management of the information, allowing the access and interchange of it without the need of a physical cable connection. The wireless transmission of voice and information has remained in constant evolution, arising new standards like Bluetooth™, Wibree™ or Zigbee™ developed under the IEEE 802.15 norm. These newest wireless technologies are oriented to systems of communication of short-medium distance and optimized for a low cost and minor consume, becoming recognized as a flexible and reliable medium for data communications across a broad range of applications due to the potential that the wireless networks presents to operate in demanding environments providing clear advantages in cost, size, power, flexibility, and distributed intelligence. About the medical applications, the remote health or telecare (also called eHealth) is getting a bigger place into the manufacturers and medical companies, in order to incorporate products for assisted living and remote monitoring of health parameteres. At this point, the IEEE 1073, Personal Health Devices Working Group, stablish the framework for these kind of applications. Particularly, the 1073.3.X describes the physical and transport layers, where the new ultra low power short range wireless technologies can play a big role, providing solutions that allow the design of products which are particularly appropriate for monitor people's health with interoperability requirements.

Keywords: WPAN, Bluetooth, Zigbee, Medical Applications, eHealth.

1 Introduction

Wireless connectivity is fast becoming recognized as a flexible and reliable medium for data communications across a broad range of applications: home and buildings applications, industrial, medical, …RF can take over where wired communication is difficult or impossible.

Also, because of the growing demand for distributed and remote sensing, data acquisition and control, the role of wireless communications only gets bigger. Sensor manufacturers are integrating RF systems in the same enclosure as their sensing devices. Data logger vendors are beginning to turn to wireless communications to enhance their products. And wireless networks are taking their place right next to traditional

M. Ulieru, P. Palensky, and R. Doursat (Eds.): IT Revolutions 2008, LNICST 11, pp. 14–20, 2009.

hardwired configurations. The industry is moving toward the implementation of networks of wireless sensors that can operate in demanding environments and provide clear advantages in cost, size, power, flexibility, and distributed intelligence.

Is clear that there are many advantages to eliminating cables in remote monitoring applications, but there are also many challenges: security, reliability, integration and power are all challenges that must be overcome before there is widespread adoption of wireless measurement systems in the consumer applications.

About medical applications, it is not limited to consumer electronics devices for simple health parameters measurement. The changing population demographics and increase in long term chronic disease will require fundamental changes in the way the world considers healthcare. Nowadays, health care is getting a bigger place into the manufacturers and medical companies, in order to incorporate products for assisted living and remote monitoring of health parameteres, where the newest wireless technologies can play a big role. At this point is necessary to talk about the integration: standards vs proprietary systems. The advantages of using a standards-based wireless network include lower costs, interchangeable products from different suppliers and a better market acceptance.

2   eHealth

The remote health or telecare -also called eHealth, is getting a bigger place into the manufacturers and medical companies, in order to incorporate products for assisted living and remote monitoring of health parameteres. At this point, the ISO 11073/IEEE 1073, Personal Health Devices Working Group, stablish the framework for these kind of applications. Particularly, the 1073.3.X describes the physical and transport layers.

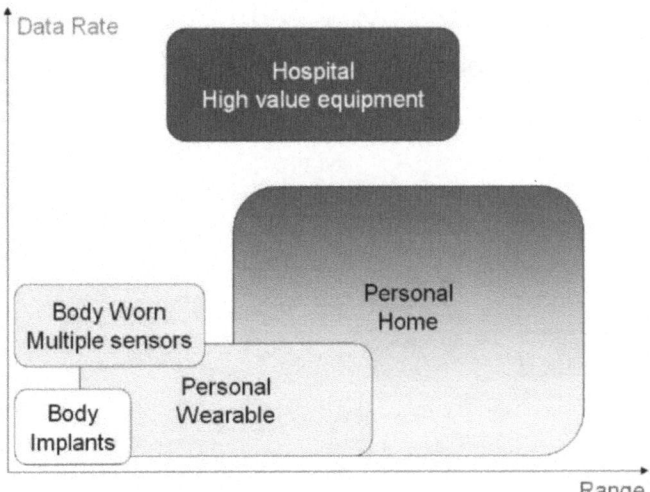

Fig. 1. The Wireless Landscape for Medical Devices

In this scenario the new ultra low power short range wireless technologies can provide solutions that allow the design of products which are particularly appropriate for monitor people's health with interoperability requirements.

eHealth is seen as central to the future of health services worldwide; it is not limited to consumer electronics devices for simple health parameters measurement. The changing population demographics and increase in long term chronic disease will require fundamental changes in the way the world considers healthcare. The current status is unsustainable as ever increasing costs burden states, industries and individuals. The clock is ticking for healthcare systems around the world. It is mandatory to develop schemes to reduce resources on the healthcare system. eHealth is recognised as the application of technology that can help to monitor people's health to reduce these costs in two ways: attendance and prevention.

Market is working with medical companies to incorporate short range wireless technologies into their products to provide a link to end to end platforms that enable these devices to talk to remote patient databases. Products particularly appropriate for Assisted Living and Dementia Care.

At its most basic level, eHealth is all about monitoring a patient's condition outside a medical institution. That can be as simple as a fall alarm, a bed occupancy or door sensor, or a remote link that connects and sends the result of a blood glucose or weight measurement to a remote database, where the data can be analysed or inspected by medical staff. The key feature is that of remote monitoring, so that a patient can have confidence in living a more normal life outside institutional care, knowing that someone is there to watch over them.

At this point is important to note that these wireless technologies are also penetrating in domotics, so it is possible to consider a home automation scenario which has not only comfort features but also health's ones due to the interoperability

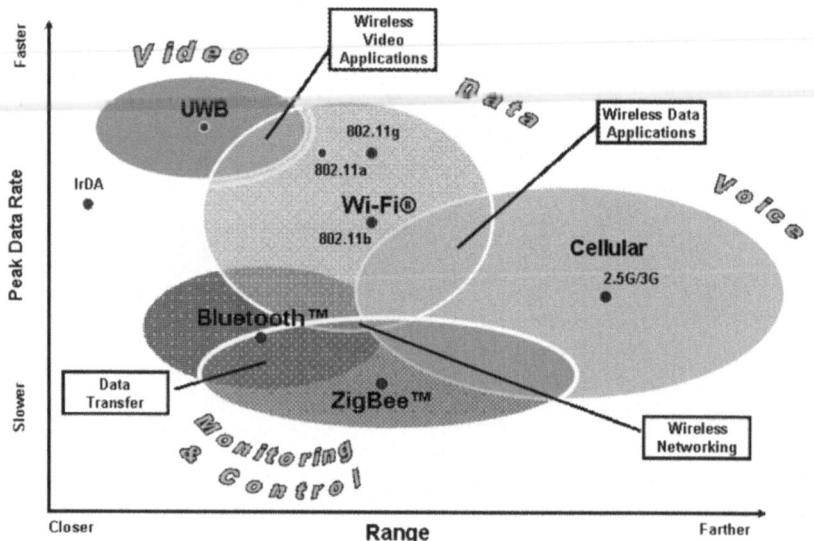

Fig. 2. Wireless technologies application areas

that underlies these standards, allowing a full integration of services in a broad concept of smart home.

We have to see how we can move from complex equipment that is manufactured in thousands –too expensive, to very cheap equipment for less sick patients that is made in tens of millions. And then how to persuade people to use them and how to manage this amount of data.

The technology for making low cost eHealth devices is largely in place. The key element that is missing is a simple[1] means of passing the patient's data to a remote database where it can be analysed. Current devices are mostly designed to be used in isolation. They need to be connected to a PC to transfer their data, which is fine for clinical use, but inappropriate for mass deployment. They also have an issue in that most use a proprietary method to record the medical information. At the point that eHealth becomes endemic and patient data needs to be managed, these formats need to be standardised.

In this case the availability of low cost medical devices that have the ability to send the data they measure to the Internet will exist through natural evolution of technology, it's unlikely to be accompanied by any centralised system for collecting and analysing it.

Three wireless standards are likely to dominate the majority of medical applications – Zigbee, Wibree/Bluetooth and Wi-Fi. Today Bluetooth and Wi-Fi dominate – Bluetooth for connections between portable devices and Wi-Fi for connections to fixed access points and IP infrastructure. Many medical researchers are currently working with ZigBee, or the 802.15.4 radio standard that underlies it.

The exception to this triumvirate of standards is the field of implantable devices. Leaving implantable devices aside, from clinical equipment through personal wellness and healthcare products, there is a diverse range of applications.

In addition to this list could appear the mobile phone network operators. They see a real benefit in offering additional services to their customers, not least to increase customer loyalty. Their 3G networks provide the wide area connections to retrieve patient data from phones and they have already spawned an innovative content management and delivery industry around their services.

Despite this diversity, four primary parameters dictate the requirements of short range wireless technology for the vast majority of medical use cases:

– the range over which the device needs to operate,
– the amount of data that needs to be transferred
– the frequency of these transfers – how often data needs to be sent, and
– the power available – typically whether it is battery or mains powered.

No standarised devices can be found nowadays, offered as consumer products. Initially these applications are likely to be centred around wellness, fitness, sports and lifestyle devices.

It is expected for 2008 that the work on the Bluetooth Medical Device Profile -that will coincide with the first standards from the IEEE 11073 Personal Health Devices Working Group, will provide a common data exchange protocol and definition of

---

[1] Simple has to be intended as normalized, standarized, interoperable.

device data formats. Together they will enable the first generation of low cost, interoperable medical devices. It's likely that the first devices we see appear will be blood pressure meters, glucose meters and weighing scales.

2.1 The IEEE 1073 Standard

The purpose is to provide real-time plug-and-play interoperability for patient-connected medical devices and facilitate the efficient exchange of vital signs and medical device data, acquired at the point-of-care, in all health care environments based on off-the-shelf technologies, scaling across a wide range of system complexities, and supporting commercially viable implementations. Next figure shows a simplified OSI layers model for this standard.

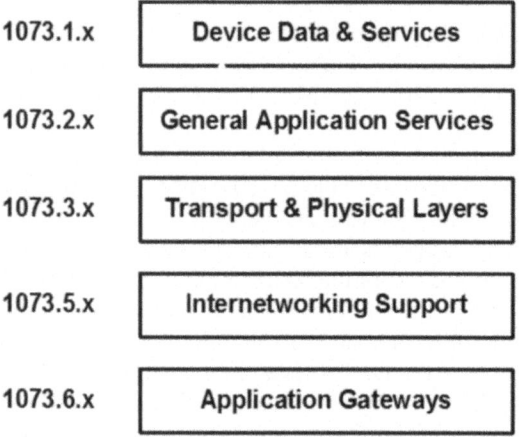

Fig. 3. IEEE 1073 simplified OSI layers

Particularly, the 1073.3.x describes the physical and transport layers. In this scenario the new ultra low power short range wireless technologies can provide solutions that allow the design of products which are particularly appropriate for monitor people's health with interoperability requirements. At this point we note the RF wireless guidelines group IEEE 1073.0.0.1 Tech Report – guidelines:

- Coordinated between medical industry, health care providers, chip manufacturers, and regulatory.
- Includes discussions on coexistence, QoS, security, and risk analysis + tech comparisons.
- Follow-on standardization projects anticipated for specific technologies (e.g., WiFi or ZigBee).

Note that the physical and transport layer appears not only in this slide or frame. It also performs the interchanging of data between different nets or standards (1073-5 and 1073-6).

3 Conclusions

The changing population demographics, the increase in life expectancy and so, long term chronic disease, will require fundamental changes in the way the world considers healthcare.

eHealth is recognised as the application of technology that can help to monitor people's health in two ways: attendance and prevention. We have to see how we can move from complex equipment that is manufactured in thousands –too expensive, to very cheap equipment for less sick patients that is made in tens of millions.

In this scenario the new ultra low power short range wireless technologies can play a big role, providing solutions that allow the design of products which are particularly appropriate for assisted living and remote monitoring of health parameteres.

Three wireless standards are likely to dominate the majority of medical applications –Zigbee, Wibree/Bluetooth and Wi-Fi. The exception to this triumvirate of standards is the field of implantable devices and in addition to this list could appear the mobile phone network operators for wide area connections.

The technology for making low cost eHealth devices is largely in place. The key element that is missing is a simple means of passing the patient's data to a remote database where it can be analysed. Also there are other challenges to achieve: security, reliability, integration and power consumption.

About the integration -standards vs proprietary systems, is expected for 2008 that the IEEE 1073 Personal Health Devices Working Group and the Bluetooth Medical Device Profile could provide a common data exchange protocol and definition of device data formats.

Finally, is important to note that these wireless technologies are also penetrating in domotics, so it is possible to consider a home automation scenario which has not only comfort features but also health's ones due to the interoperability that underlies these standards, allowing a full integration of services in a broad concept of smart home.

References

1. Choi, S.-H., Kim, B.-K., Park, J., Kang, C.-H., Eom, D.-S.: An Implementation of Wireless Sensor Network for Security System using Bluetooth. IEEE Transactions on Consumer Electronics 50(1), 236–244 (2004)
2. Kammer, D.: Bluetooth Application Developer's Guide: The Short Range Interconnect Solution. Syngress Publishing Inc. (2002)
3. Bellido Outeiriño, F.J., de la Cruz Fernández, J.L., Torres Roldán, M., Gistas Peyrona, J.A.: Comunicación Inalámbrica con Bluetooth, Revista Técnica Industrial, pp. 18-23 (October 2004) ISSN: 0040-1838
4. Bellido Outeiriño, F.J.: Aplicación de los nuevos sistemas de comunicación inalámbrica a sistemas de adquisición y transmisión automática de la información. Doctoral Thesis. University of Cordoba, Spain (2007)
5. Bellido Outeiriño, F.J., de la Cruz Fernández, J.L., Torres Roldán, M., Moreno Muñoz, A.: Wireless technology applied to stimulation systems for auditory deficit children. In: Proceedings of the 12th IEEE International Symposium on Consumer Electronics (ISCE 2008), Vilamoura-Portugal (2008)

6. Moreno-Munoz, A., Flores, J.M., Bellido, F., de la Rosa, J.J.G.: Integration of Power Quality into distribution automation through the use of AMR. In: Proceedings of the IEEE International Symposium on Industrial Electronics ISIE 2008, Cambridge, UK, pp. 1633–1638 (2008) ISBN: 978-1-4244-1665-3
7. Hunn, N.: Wireless in Medical Technology. Ezurio Ltd. Rev. (2007)
8. Cooper, T.: ISO/IEEE 11073 MDI 2 Point-of-Care Medical Device Communication: Overview & Status Update. American Telemedicine Association. Breakthrough Solutions (2005)
9. Galarraga, M., Serrano, L.: Interoperabilidad de Dispositivos Médicos. IEEE 1073, Universidad Pública de Navarra (Spain) (2006)
10. ZigBee Alliance. Technical Documents (September 2008), http://www.zigbee.org
11. Bluetooth SIG. Technical Documents (September 2008), http://www.bluetooth.org
12. Wibree: Technical Documents (September 2008), http://www.wibree.com

Implementation of Virtualization Oriented Architecture: A Healthcare Industry Case Study

G Subrahmanya VRK Rao, Jinka Parthasarathi, Sundararaman Karthik,
GVN Appa Rao, and Suresh Ganesan

Cognizant Technology Solutions
Chennai, India
{subrahmanyavrk.rao,parthasarathi.jinka,karthik.sundararaman,
apparao.gvn,ganesan.suresh}@cognizant.com

Abstract. This paper presents a Virtualization Oriented Architecture (VOA) and an implementation of VOA for Hridaya - a Telemedicine initiative. Hadoop Compute cloud was established at our labs and jobs which require a massive computing capability such as ECG signal analysis were submitted and the study is presented in this current paper. VOA takes advantage of inexpensive community PCs and provides added advantages such as Fault Tolerance, Scalability, Performance, High Availability.

Keywords: Virtualization Oriented Architecture, Telemedicine, Technology Convergence, Cloud Computing.

1 Introduction

Distributed processing and scaling continues to gain prominence across the Information Technology (IT) infrastructure landscape. Architectural approaches which adapt this foundation offer inherent advantages, including the potential for price-performance economic benefits and greater flexibility. To address this challenge, organizations are exploring software that operates well across distributed (scale out) and centralized (scale up) resources, and offers a consistent application experience and approach across a number of heterogeneous resource types. Virtualization is one such concept that has got a set of technologies which serve the proposed need and is typically, "the provision of an abstraction between a user and a physical resource in a way that preserves for the user the illusion that he or she could actually be interacting directly with the physical resource"[1].

Virtualization Oriented Architecture (VOA) calls to mind the use of virtualization to make service oriented applications more durable and versatile. VOA is neither a technology nor a technology standard. Instead, it represents a technology-independent, high-level concept that provides a set of architectural patterns or a blueprint [2]. These patterns are focused on the componentization and composition of the enterprise application in such a way that the components are created and exposed as Virtual Services/ Jobs which run on a virtualized infrastructure through a Virtual engine of an Enterprise. These Virtual Services / Jobs are not only technically independent but also have a direct relationship to the business process and are adapted for the Virtual Infrastructure.

M. Ulieru, P. Palensky, and R. Doursat (Eds.): IT Revolutions 2008, LNICST 11, pp. 21–27, 2009.
© ICST Institute for Computer Sciences, Social-Informatics and Telecommunications Engineering 2009

Telemedicine is a branch of healthcare industry which assists the people in remote places to communicate information about their health to an expert physician. The information is mostly in the form of biomedical signals and images. Large storage devices and expert decision support systems can assist the physicians in storing and analyzing these medical records. Currently healthcare provision for people living in remote places is facing many problems like accessibility to the geographic location, time required for the physicians to analyze and store massive medical records and provide the necessary treatment etc. These problems could be overcome if one adopts high performance computing and communication technology which can potentially enhance patient care by providing an easy way to store and retrieve massive amount of data and transfer medical records quickly to a nearby hospital for an expert opinion and facilitate subsequent treatments.

Virtualization has the potential to make data storage and retrieval easier, cost efficient and less time intensive for healthcare industries [3]. A combination of advanced IT techniques such as Virtualization, Cloud computing, Software As A Service (SaaS) and wireless communication technology can be very useful for both the patients (who want to have easy access to the physician) and healthcare providers (who require an efficient storage, retrieval and analytical system which can save money and time).

The current paper presents ongoing work at our Research Labs. Hadoop is a software framework that supports data intensive distributed applications. We have established a Hadoop [4] Computing Cloud across inexpensive community PCs. We have developed a Virtual Engine Software. We have chosen a Compute Intensive task/job from Healthcare Industry such as ECG Signal Analysis and we submitted these compute intensive jobs to the Hadoop Cloud through our Virtual Engine Software and the current paper presents the same study.

2   Virtualization Oriented Architecture

Fig. 1 illustrates the Component View of VOA and Fig. 2 illustrates the Operational View of VOA. The current section presents descriptions of various components of VOA.

Application FrontEnds are the essential ingredients of VOA. Although they are not services themselves, they initiate all business processes and ultimately receive their results. Typical Application FrontEnds are Graphical User Interfaces (GUIs) or a Batch Process.

A *Virtual Service/Job* (Fig. 3) is the implementation of a Service that adapts to the Virtual Infrastructure/ Hardware .

Virtual Engine connects all participants of a VOA i.e., Virtual Services/Jobs, Infrastructure/Hardware and Application FrontEnds.

Virtual Service Repository captures all essential information about the Services so that the enterprise can be aware of its existence and capabilities. The Virtual Service Repository contains the service contract and additional information such asPhysicalLocation, Information about the Provider, Contact Details, Technical Constraints, Security issues, and available Service Levels.

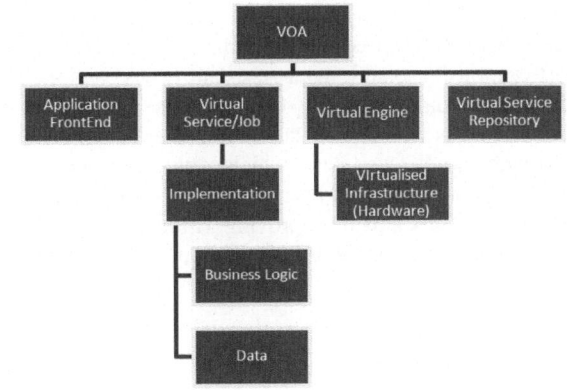

Fig. 1. Component View of VOA

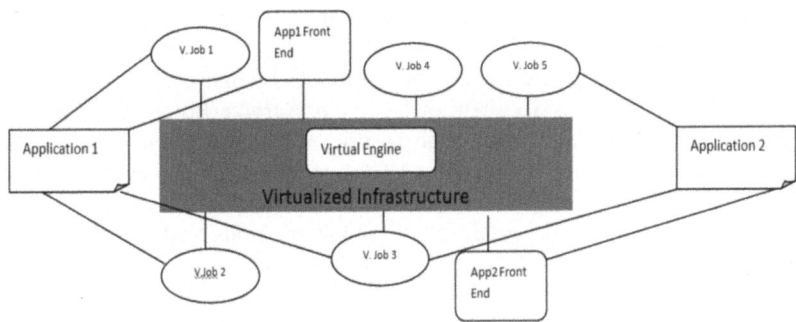

Fig. 2. Operational View of VOA

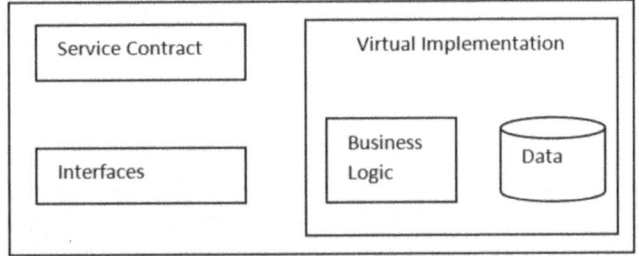

Fig. 3. Illustration of a Virtual Service / Job

3   Hridaya – A Telemedicine Application

'Hridaya' is a Tele-Medicine initiative based on a Convergence of Grid, Web 2.0, WiMax Technologies and was presented by the authors elsewhere [5].Cardiovascular disease

(CVD) is a major cause of death among humans. Rehabilitation after survival from a CVD is a long term process that includes frequent hospital visits, which could be avoided by the use of Hridaya Software on Personal Digital Assistant (PDA) and mobile phones. These devices along with Hridaya could be used by the patients to report about their health to the physicians on a schedule basis who in turn can review their health conditions. In addition, Hridaya is equipped with a knowledge base, which could be used by the patients to know more about CVD, an interface powered by technologies underlying Web 2.0 which has several advantages [6]. A VOA aids in improving Throughput, Availability, Scalability, Fault Tolerance and reduces the complexity of infrastructure management for the vendor providing Hridaya.

4 VOA Implementation of Hridaya

Usually, Hridaya receives Bio-Medical Signal from the user through User Interface and processes the same and do provide report on Body condition of the User/Patient and the report related graph. VOA implementation of Hridaya involves a Virtual Engine, which indeed runs a 'Hridaya' Service i.e., the 'core analysis' function as a 'service'. This enables processing of the Compute-Intensive Bio-Medical signals from the End User (such as a Patient or Doctor or Service Provider). The 'service' processes the incoming signals through a series of computations and returns the output to the Decision Support System (A Component of Hridaya), which analyzes the signal and returns its output/decision in turn to the Physician/End User; which would be a report on the body condition of the patient along with the related graphs. The job could be submitted to the system through a portal User Interface, as illustrated in Fig. 5.

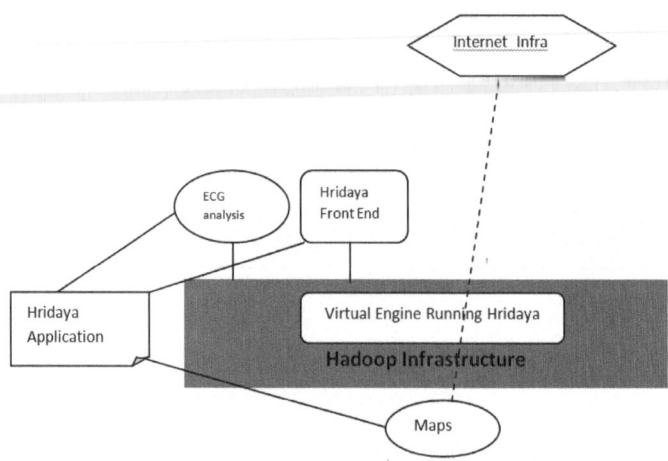

Fig. 4. VOA Implementation of Hridaya

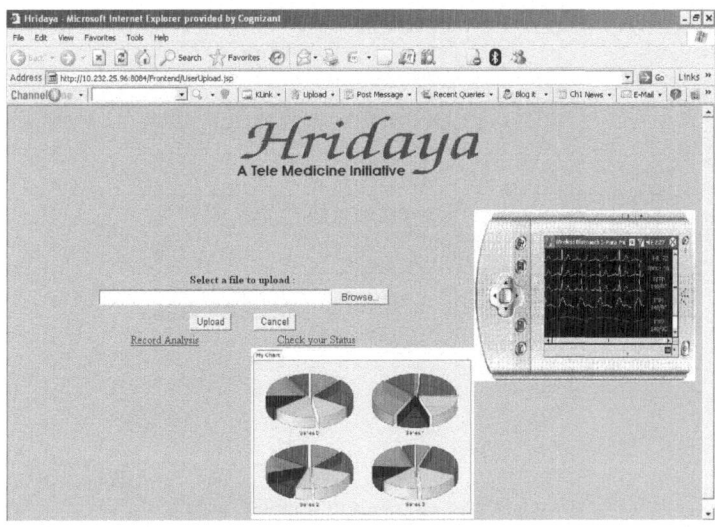

Fig. 5. Hridaya – User Interface

The service receives the signal from the user. Since the service is implemented over the Internet, we use SOAP (Simple Object Access Protocol) to attain interoperability with the user interface. SOAP is a simple XML based protocol to let applications exchange information over HTTP. The XML file acts as a carrier to transfer the signal from the user to the service. When the service receives this XML file it extracts the signal from the XML file and stores the file in the Hadoop Cluster infrastructure as shown in Fig. 4. The signals are usually stored in files. This data will be split into number of Blocks and distributed over the Hadoop cluster using Hadoop Distributed File System (HDFS). The Hadoop cluster [4] contains Namenode and Jobtracker that act as masters; Datanode and Tasktracker that act as slaves. Namenode provides information about the number of blocks obtained, number of nodes and the number of Blocks present in various nodes. Datanodes provides the information about how the blocks are received from various nodes to the corresponding node and how their storage is managed. Jobtracker provides identity for all the tasks and chooses the tasks to be performed by the particular Tasktracker. It provides information about the status of the task and saves the output in the HDFS. Tasktracker is a node containing the Block, performs Map-Reduce task assigned by the JobTracker running the Virtual Engine as shown in Fig. 4. The Map task does a Fast Fourier Transform (FFT) on the input signal Block and runs on the TaskTrackers. These map tasks generate an intermediate output which is again, an another signal. The map output is now sent to the Reducer which develops two graphs, one for the input signal as shown in Fig. 6(a) and one for the output signal in Fig. 6(b). These two graphs are part of the output that has to be sent to the user. The Reducer also performs the decision support system which correlates the signal with Pre-Recorded signals and gives a condition report which along with the graphs is sent

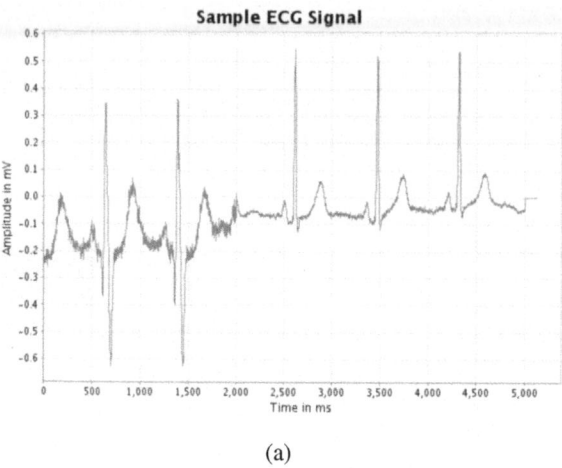

(a)

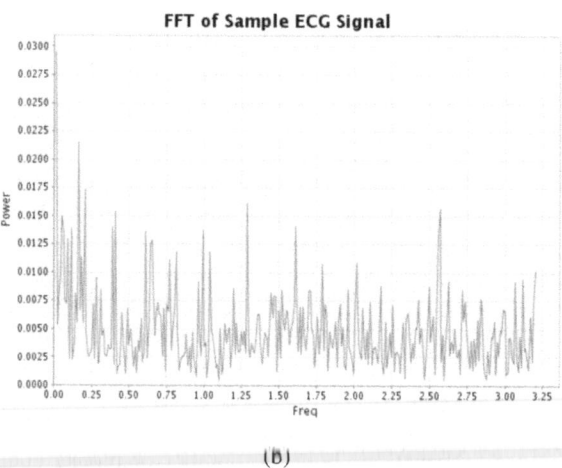

(b)

Fig. 6. Hridaya 'Analysis' - In Action

back to the HDFS. This report if abnormal can alert the physician to review the ECG with high priority and take necessary steps to treat the patient. The Web Service will again receive the output from the HDFS and attach the output files and data again to a XML file. The User Interface will now retrieve the graphs and the report from the XML file and display the data and the graphs required by the User as and when requires.

5 Conclusion and Future Work

VOA which is based on Virtualization Technology will leverage the inherent advantages of Cloud Computing and provide advantages such as Fault Tolerance, Scalability, Performance, High Availability. Our Research Team is working on other related use-cases across various Industry Sectors. We are also working on Maintenance and Performance etc. issues of the Hridaya system.

References

1. http://www.acmqueue.org
2. SOA Glossary of Key Concepts and Definitions,
   http://hssp-infrastructure.wikispaces.com/space/showimage/
   Microsoft+Word+-+HSSP+SOA+Key+Concepts+and+Definitiions+
   v0.3.pdf
3. http://t1d.www-03.cacheibm.com/industries/healthcare/doc/
   content/bin/Virtualization_UPMC_profile_IDC_1_07.pdf
4. http://www.hadoop.com
5. Subrahmanya, G., Rao, V.R.K., Parthasarathi, J.: Convergence of Web2.0 and Grid accelerating SaaS – An Insight. In: International conference on Convergence Information Technology, Korea (2007)
6. Karthik, S., Parthasarathi, J., Subrahmanya, G., Rao, V.R.K., Appa Rao, G.V.N.: Hridaya –
   A telemedicine initiative for cardiovascular through convergence of Grid, Web2.0 and
   SaaS, Ambient Technologies for Diagnosing and Monitoring Chronic Patients Tampere. In:
   2nd International Conference on Pervasive Computing Technologies, Finland, January 29
   (2008)

Location Tracking Strategy Indicating Sufferers' Positions under Disaster in the Buildings

Min-Hwan Ok

Korea Railroad Research Institute
360-1 Woulam, Uiwang, Gyeonggi, Korea 437-757
mhok@krri.re.kr

Abstract. The advancement of location-based services now covers indoor location positioning. Under disaster in the building, the sufferer might faint, be wounded, or enclosed by structures in the dark. In the cases, the sufferer could not let the relief team know her position in the building. The LBS server provides location tracking or positioning of her device for quick relief. In the service UltraWideBand is used by its good penetrability. In dead-reckoning operation that the device is lost on the sensor network, the relief team traces logged profiles of location tracks. The strategy regards the privacy concerns.

Keywords: Location Based Service, Ultra Wide Band, Disaster and Relief, RFID, Sensor Network.

1 Overview

Location-based Services, LBS, have been developed from outdoor ones to one for indoor ones. Most of outdoor LBS employed Global Positioning System, GPS, and equipped on vehicles and vessels. From the conveyance-oriented application of vast ranges, the issues of LBS have covered to the human-oriented application of narrow domains nowadays in the paradigm of ubiquitous networking. During indoor life, there have been many disasters including collapse, fire, and flood. Under the disaster, the location tracking or positioning of the sufferers is essential for quick relief.

Indoor LBSs are distinguished by several wireless communication technologies including WiFi, Infrared, Supersonic, Bluetooth, RFID and UitraWideBand. In WiFi, the device measures the signal strength of RF signal. Microsoft RADAR[1], Ekahau[2], Intel Placelab[3] adopted WLAN. AT&T Lab.'s Active Badge[4] uses Infrared, and Active Bat[5] uses Supersonic technologies. Bluetooth exploits RSSI(received signal strength intensity). SpotON[6] uses RFID, and Ubisense[7] Ubitag uses UWB. Among the wireless communications, UWB has appropriate characteristics such as low consumption, wide spectrum frequencies, high resolution and most importantly, good penetrability for indicating sufferers' position behind non-metal structures. Furthermore the signal reaches near $50m$, which is considered to be a suitable distance to LBS in the buildings.

While wireless communication is alive, indicating sufferers' locations is positive, but while it is not, alternative-indicating method should be necessarily prepared. For

M. Ulieru, P. Palensky, and R. Doursat (Eds.): IT Revolutions 2008, LNICST 11, pp. 28–31, 2009.
© ICST Institute for Computer Sciences, Social-Informatics and Telecommunications Engineering 2009

dead- reckoning operation, one alternative method is logging profiles of location tracking. A LBS server logs a profile of the device's movement in the building only. The device's location is traced along the logged profile in dead-reckoning operation.

2   Location Tracking to Indicate Sufferers' Positions

The calculation task of the sufferer's position could be processed in ether the device or the LBS server. If the task is processed in the device, the device may consume its power earlier and it is not desirable situation under disaster. Thus the LBS server takes calculation tasks of the devices. The device merely transmits its beacon signal at regular intervals. Fig 1. depicts the location positioning of the devices.

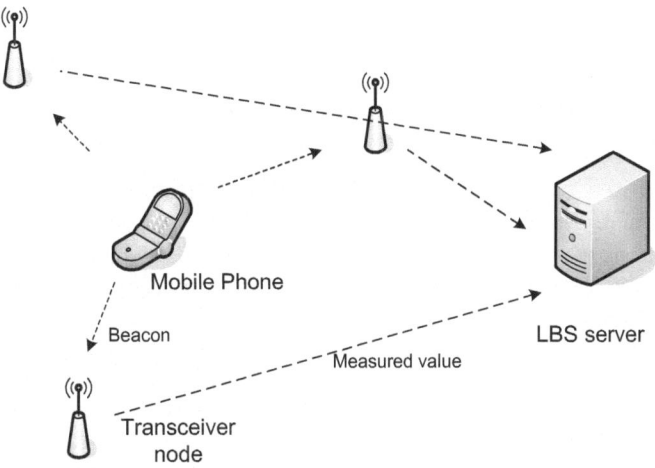

Fig. 1. The LBS server calculates the positions of the devices

Under disaster, the relief team moves with the location tracking devices, *Indicator PDAs*, to find the sufferers out. The Indicator PDA calculates its position not to burden the LBS server important for finding out the sufferers. In this situation, the transceiver also transmits their beacon signals to send their status. For the locations of some transceiver nodes might be changed by disaster, more than 3 transceiver nodes are used in location positioning of multiple trilaterations with multiple sets of 3 nodes. The Indicator PDA sends its calculated position for logging of their positions. Fig. 2 depicts the location positioning of the Indicator PDAs.

The positions of other devices near are indicated on the Indicator PDAs. The console of the LBS server becomes a part of the command post. All the active devices and Indicator PDAs are shown on the console screen. In the case the LBS server is off-line, the indicator PDA should process the calculation task of its position together with other devices' positions transmitting their beacon signals. Therefore the team carries multiple Indicator PDAs.

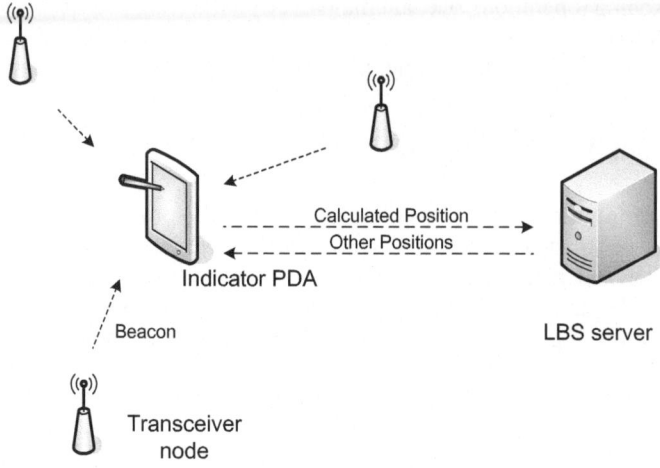

Fig. 2. The Indicator PDA calculates its position and sends it to the LBS server

3   Location Tracing in Dead-Reckoning Operation

Under disaster, there are many cases the device does not work. The sufferer may be located behind a metal structure. The required transceiver node may malfunction or not work within domain. The device could be lost or broken, and its battery could run out. For the cases, the LBS server logs a profile of the device's movement in the building only. The logged profile contains the ID assigned instantly during residence in the building, and positions along times. This profile is created on entrance, and deleted on exit of the building, for privacy concerns.

If there are IDs assigned but do not appear on the sensor network, the LBS server displays the IDs and shows their tracks and final locations. The relief team would deliver more transceiver nodes around the final locations. For the safety of logged profiles duplicating the LBS server is required.

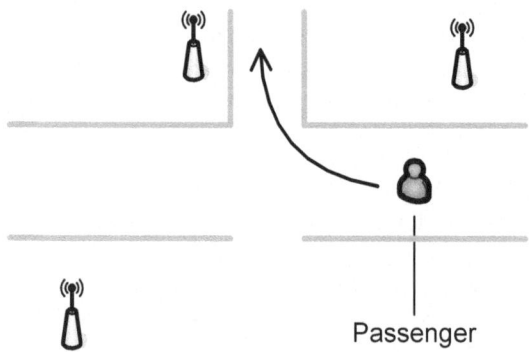

Fig. 3. Location tracking for use in dead-reckoning operation

For the people who don't carry any device, an RFID solution may be practical; allocating RFID tags at entrance and recollecting them at exit of the building. The transceiver nodes have RFID readers and sends location tracks to the LBS server as shown in Fig. 3. Those people are only traced by logged profiles. For example, the passengers of railway stations receive tickets with RFID tags. On departure, the passengers have exited the station and on arrival, the passengers have entered another station. They return their tickets with RFID tags and then disappear from the sensor network of the station.

It is crucial how wireless communications could be well transmitted under disaster including fire, flood, and collapse. Although UWB has a characteristic of good penetrability, there may be many obstacles to transmissions under disaster. The LBS system itself could be damaged, and thus a group of Indicator PDAs could be the only counted on. For this situation, the Indicator PDAs should form an ad-hoc network replacing the LBS server off-line.

References

1. Bahl, P., Padmanabhan, V.N.: RADAR, An In-Building RF-Based User Location and Tracking System. In: IEEE INFOCOM 2000, vol. 2, pp. 775–784. IEEE Press, Los Alamitos (2000)
2. Ekahau RTLS, http://www.ekahau.com
3. Place Lab, http://www.placelab.org
4. Harter, A., Hopper, A.: A Distributed Location System for the Active Office. IEEE Netw. 8(1) (1994)
5. Hightower, J., Borriello, G.: Location systems for ubiquitous computing. IEEE Comp. 34(8), 57–66 (2001)
6. Hightower, J., Want, R., Borriello, G.: Spoton: An Indooe 3D Location Sensing Technology Based on RF Signal Strength, Technical Report, University of Washington (2000)
7. Ubisense platform, http://www.ubisense.org

Measuring Cognition Levels in Collaborative Processes for Software Engineering Code Inspections

David A. McMeekin, Brian R. von Konsky, Elizabeth Chang, and David J.A Cooper

Digital Ecosystems and Business Intelligence Institute, Curtin University of Technology,
Enterprise Uni 4, De' Laeter Way, Technology Park, Bentley WA 6102, Australia
{d.mcmeekin,b.vonkonsky,e.chang,david.cooper}@curtin.edu.au

Abstract. This paper demonstrates that different software code inspection techniques have the potential to improve developer understanding of code being inspected to varying extents. This suggests that some code inspection techniques may be superior to others with respect to improving the efficacy of future inspections, harnessing collective wisdom, and extending team knowledge and networked intelligence. In particular, this paper reports results from a study of novice developers' cognitive development during a software inspection training exercise. We found that developers who performed a code inspection prior to modification tended to operate at higher cognitive levels beginning very early in the modification exercise. Those who had not performed an inspection tended to operate at lower cognitive levels for longer periods of time. Results highlight the importance of code inspections in increasing developers' understanding of a software system. We believe collaboration between academia and industry in studies such as these would benefit the three major stakeholders: academia, industry and graduates.

Keywords: Collaboration, Collective effort, Software inspections, Bloom's taxonomy, Programmer comprehension, Cognition development.

1 Introduction

Industry-based software inspection processes are normally used to detect software defects. However, they can also have an impact on a developer's understanding of the system being inspected, with the potential to improve team cognition levels and the effectiveness of future collaborative inspection exercises.

A software inspection was used as a training exercise prior to developers adding functionality to code. During both the training and coding exercises, the software developers' cognitive levels were measured using the Context-Aware Analysis Schema [15]. One group of developers had not seen the code prior to adding functionality. The other group of developers had inspected the code immediately prior to adding the functionality using one of three inspection-reading techniques: Ad hoc reading, Abstraction Driven Reading or Checklist-Based Reading.

M. Ulieru, P. Palensky, and R. Doursat (Eds.): IT Revolutions 2008, LNICST 11, pp. 32–43, 2009.
© ICST Institute for Computer Sciences, Social-Informatics and Telecommunications Engineering 2009

2 Industry Practice

Software Inspections are a practical methodology widely used in the ICT industry. They are a tried, tested, and effective method for the removal of defects from software early in the development life cycle [11]. Additionally, software inspections assist inspectors in developing greater insight and understanding into the artefact being inspected [16]. IDE's such as Netbeans, Eclipse, Xcode and Visual Studio, with their auto completion functions for example, has meant developers are warned of many potential errors before finishing writing the line of code. For example, when calling a function, the auto completion will display the method signature. This assists the developer to order the parameters correctly.

Reading software artefacts is an essential practice for producing high-quality software during a product's development and maintenance life cycle [2]. Inspection techniques/reading strategies are usually linked with verification and validation of software artefacts. Applying inspections to raise cognition levels and reading skills is an area that has not been well researched as an additional possible benefit of software inspections.

Traditionally software inspections are a collaborative task, typically comprising four inspectors. The first person is the moderator, who presides over and manages the team inspection process. The second is the designer responsible for the design of the code in question. The third is the implementer who translated the design document into code. The final participant is the tester who was responsible for writing and executing the test cases.

2.1 Practical Software Inspection Methods

Software inspections are implemented early in the development process to detect defects in the inspected artefact [11], and offer developers a structured method to examine software artefacts for defects [9]. Software inspections and their success in detecting defects is a well-researched area in software engineering [23].

Performing a code inspection prior to modifying the code has been shown to improve a developer's ability to carry out the required changes [19]. The inspection techniques tested in this study were: ad hoc, Abstraction-Driven Reading (ADR), and Checklist-Based Reading (CBR).

The ad hoc technique is understood to be the simplest inspection technique to use. No formal methodology is used when applying this method. The inspector is expected to thoroughly inspect the artefact using his/her personal experience as the guide [16].

This method's strength lies in giving the greatest freedom to the inspector as to how they execute the inspection [9]. Its greatest weakness, correspondingly, is uncovered by novice developers, who lack the necessary experience to effectively apply it [16].

The Abstraction-Driven Reading (ADR) technique was created in response to the delocalisation challenge Object-Orientation introduced to traditional inspections [8] [9]. The inspector reads code in a systematic way, writing natural language abstract specifications about each method and class. While reading each method, calls to delocalised code are followed and the invoked code is also inspected. As the inspector systematically executes these tasks they also compile a list of detected defects.

A strength of this technique is the requirement for inspectors to develop reusable natural language descriptions of the inspected code. However, this comes with concomitant costs in time, and can be overwhelming for the inspector to attempt to grasp an understanding of the whole program.

Checklist-Based Reading inspections were formally introduced by Fagan [11] and are considered the standard inspection method used by software organisations today [16]. The inspector has a series of questions that guide their reading. The questions should be derived from historical data from within the organisation identifying defects detected in previous systems [12][13].

Each question on the list requires a yes or no answer. A yes answer implies no defect in the code at that location. A no answer indicates the possibility of a defect there and necessitates a closer examination of the code.

A strength of this technique is that the checklist is a product of prior inspections and captures organizational history with respect to the cause of prior defects. A weakness is that it is a highly structured process that can restrict inspectors from reading the code in a more natural manner.

3 Bloom's Taxonomy for Educational Objectives

3.1 Bloom's Cognitive Development

Bloom's taxonomy is a well-established categorisation of six different cognitive levels potentially demonstrated during learning [1] [4]. The categories range from the lowest to the highest level of cognitive learning. The classification is widely used in education systems throughout the world.

Each category, cited from [1], is listed and briefly described below, with an example of how each might be translated in a programmer's context:

Knowledge: "retrieving relevant knowledge from long-term memory." For the programmer this may be the specific recalling of an if-then-else statement.

Comprehension: "construct meaning from instructional messages, including oral, written, and graphic communication." For the programmer, summarising a method or code fragment.

Application: "carry out or use a procedure in the given situation." Demonstrated when the programmer makes a change in the code.

Analysis: "break material into constituent parts." Where the programmer describes a method or field's operation and role within the wider system.

Evaluation: "make judgements on criteria and standards." Here the programmer makes a judgement on the correctness or incorrectness of a part of the program.

Synthesis: "re-organise elements into a new pattern or structure." The programmer creates a new class, successfully integrating it into the wider system.

3.2 Industry-Based Context-Aware Schema Using Bloom's Taxonomy

Bloom's taxonomy has been used in many software engineering studies to examine developers' comprehension levels during different tasks [6][15][25][26][27]. Kelly and Buckley [15] developed a Context-Aware Schema for use with the taxonomy. The schema requires developers to "think-aloud" as they perform the different tasks required of them. Think-aloud is a process in which the participant verbalises thoughts and actions while carrying out the task [10].

The think-aloud data is recorded, transcribed and broken down into sentences or utterances. Each sentence or utterance is then categorised into a level within the taxonomy to identify the cognitive level at which the developer was operating. Each utterance is categorised upon both its content and the previous two utterances. This enables the utterance to be categorised within its applied context.

The original Context-Aware Schema [15] omitted the synthesis level, as their study was carried out in a maintenance environment. In our study we have introduced the synthesis category. This was because developers were required to add new functionality to the code. This new functionality was not fixing defects but rather extending the program to perform a new task.

4   Methodology for Collective Academic and Industry Learning

To investigate understanding arising as a result of the various code inspection strategies, a study was conducted in which novice developers were required to add new functionality to an existing software system. During this process their cognition levels were measured using the Context-Aware Schema [15].

The software system was the game of Battleship. It was a text-based implementation written in Java and contained seven classes in total. The Board.java class required new functionality; the ability to place ships in a diagonal down to the right manner. Participants were given 30 minutes to add this new functionality, and were required to think-aloud for the task's duration. When their time was up, participants stopped regardless of whether they had completed the task or not.

The study was advertised on the university campus and participants took part in their own time. No compensation was paid to participants and they were informed that participation had no influence whatsoever on their marks/grades in courses they were currently undertaking. Participants were required to be final year undergraduate studying Software Engineering, Computer Science or Information Technology.

Participants were provided with the following artefacts for the modification task:

- a natural language description of the system,
- a class diagram,
- the Board.java file (without defects),
- access to the other Java code in the system,
- a natural language modification request, and
- access to the Java APIs. All artefacts were online.

Prior to adding the functionality, four participant groups were established and individual members from three of the groups performed a 30-minute code inspection on the Board.java class, searching for defects. The think-aloud data was also collected from the inspection task. Group One performed an ad hoc inspection, Group Two performed an ADR inspection, Group Three performed a CBR inspection, and Group Four did not perform an inspection. Participants performed two small training exercises prior to participating in the study. The first exercise involved using the assigned inspection technique. The second exercise detailed how to think-aloud.

Participants were informed that the defects seeded in the code were not syntax–related, as the code compiled and executed. They were searching for defects that would cause the system either to fail or produce incorrect output.

Research of this nature, based on empirical studies, is subject to internal and external validity threats. The first internal threat to this study was the selection threat. Selection threat is where participants are stacked to produce favourable results. To limit this threat, the study was advertised on campus, all final year students who asked to participate were admitted to the study, and all participants were randomly assigned to the different inspection technique groups.

The second internal threat, as with many software engineering studies, was variation in participant experience. In considering this threat, demographics were collected from participants in order to monitor discrepancies that may have arisen within the results from this.

The external validity threat in this study was the sample size. There is significant overhead involved when using the think-aloud method. The data must be collected, collated, transcribed, broken into utterances, and analysed. The sample size was kept to 20 as the research was attempting to identify any emerging trends within the data that could be pursued with larger data sets in the future. It must be noted that even with small sample sizes, although difficult to generalise to a larger body, significant differences may still be identified [20] warranting continued research.

5   Results on Effective Bloom's Cognition Development and Software Skill Training

Table 1 displays an utterance example from each of the 6 different categories of Bloom's taxonomy. The seventh category ('Graph Number 0) shown in Table 1 is Uncoded, and is not part of the taxonomy. Utterances in this category were either unintelligible or unrelated to the task at hand, such as talking on the mobile phone during the study.

Table 1. Example of utterances

Graph Number	Bloom's level	Utterance Example
1	Knowledge	"while ship not sunk"
2	Comprehension	"this is a one to many relationship"
3	Application	"we need to cater for a new direction"
4	Analysis	"this is externally controlled"
5	Evaluation	"the call here is incorrect"
6	Creation	"creating a new method"
0	*Uncoded*	*"what food will we need for tonight"*

Figures 1, 2, 3 and 4 graph four different participants' utterances in order of their occurrence, categorised into the appropriate cognitive level using the Context-Aware Schema. The X-axis shows the order of the utterances and the Y-axis represents utterance's cognitive level.

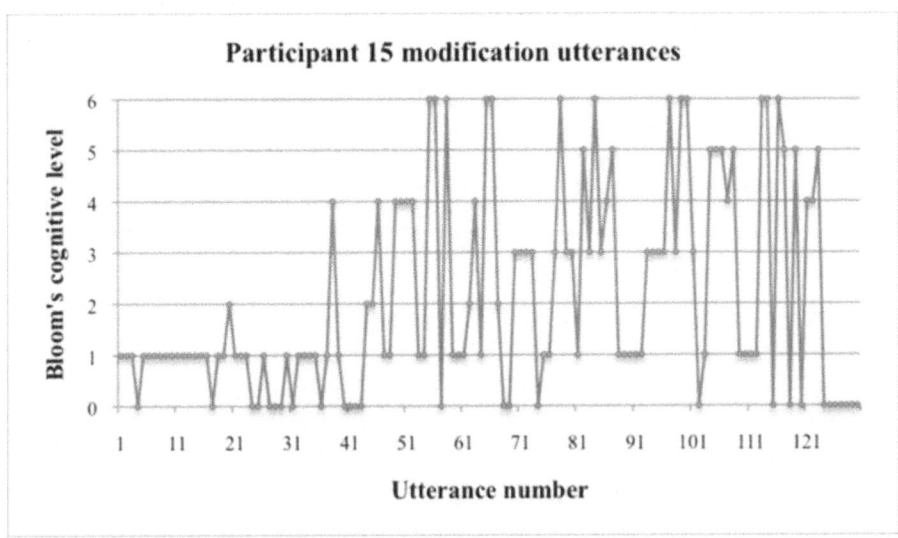

Fig. 1. Participant 15's modification utterances. Performed no inspection.

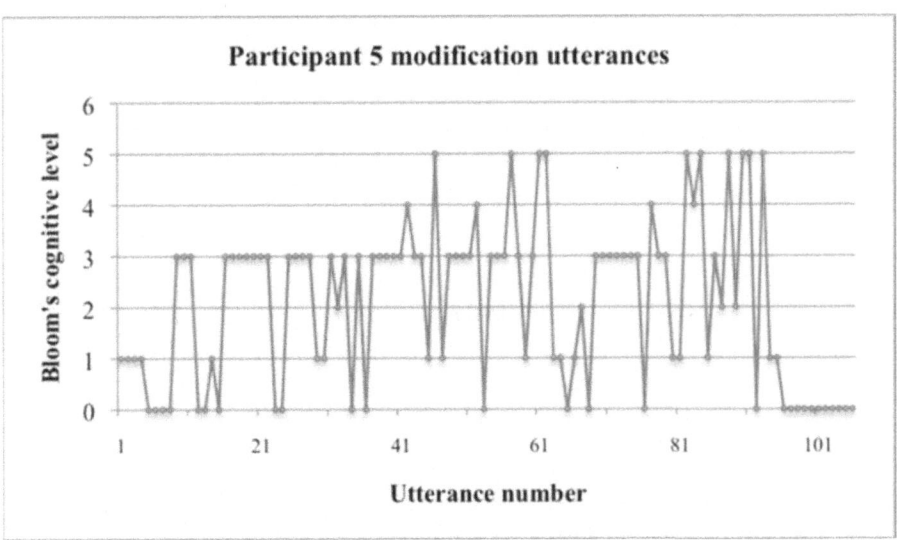

Fig. 2. Participant 5's modification utterances. Performed a CBR inspection.

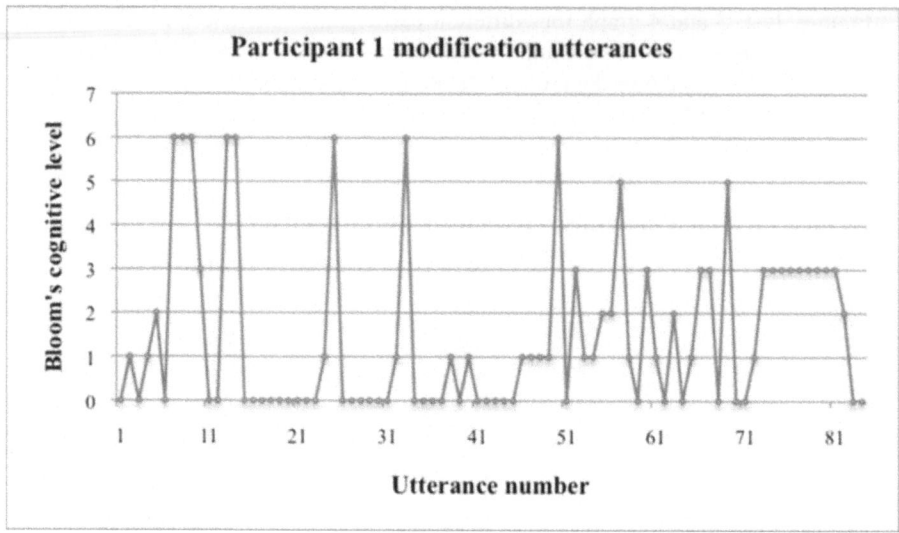

Fig. 3. Participant 1's modification utterances. Performed an ad hoc inspection.

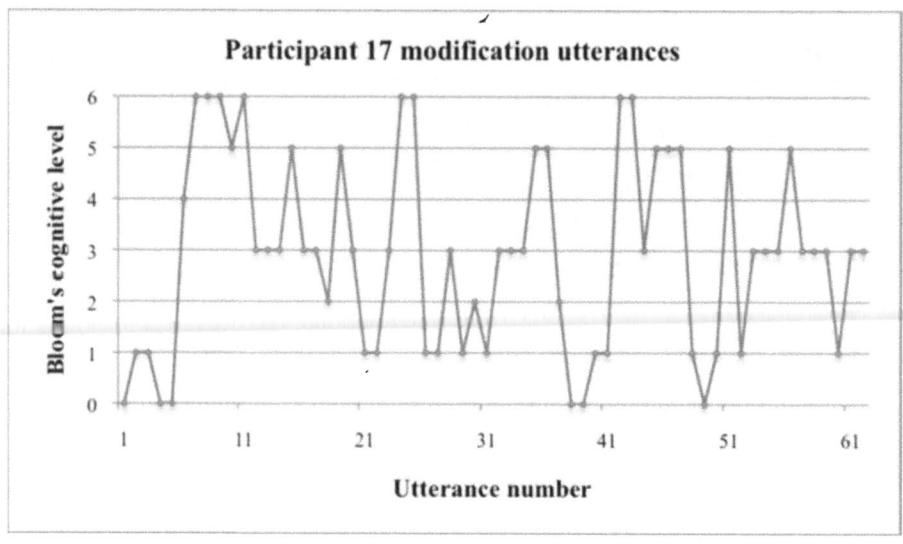

Fig. 4. Participant 17's modification utterances. Performed an ADR inspection.

Figures 5 and 6 are graphed in a similar manner, but the utterances are from the inspection each participant performed. Due to the large number of graphs, only samples of the participants have been displayed in this paper.

Figure 1 demonstrates almost 50% of participant 15's modification utterances were in the lowest cognitive level, Knowledge. The participant was unfamiliar with the code prior to receiving the modification request. Hence, in the 30 minutes given for

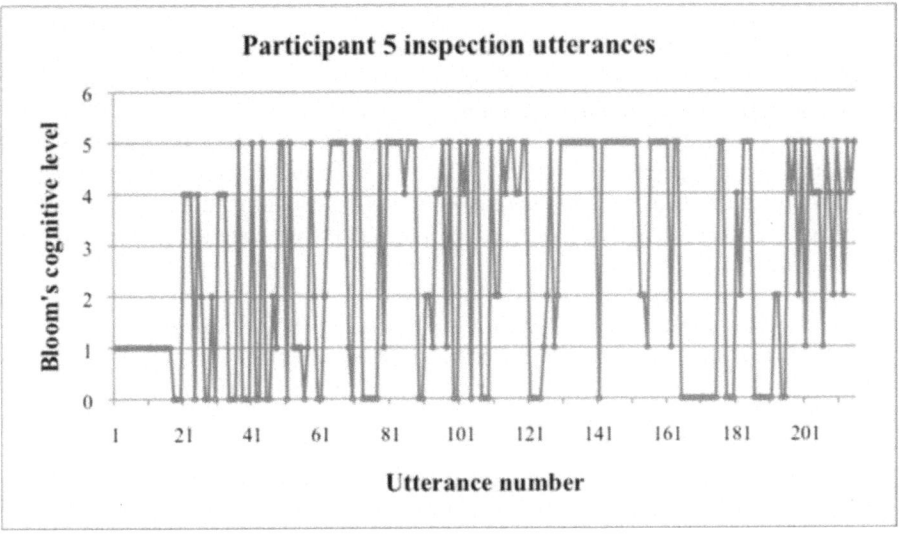

Fig. 5. Participant 5's utterances during inspection. Performed a CBR inspection.

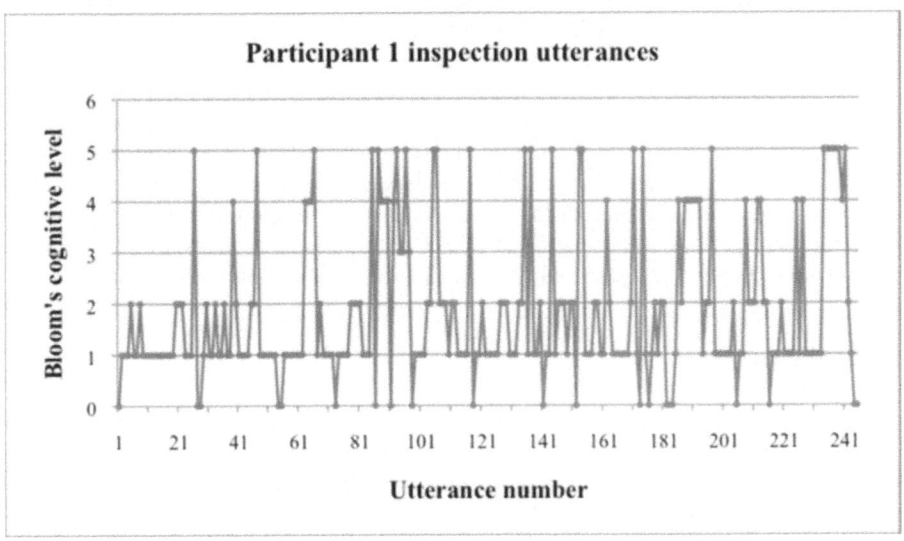

Fig. 6. Participant 1's inspection utterances. Performed an ad hoc inspection.

the task, they needed to familiarise themselves with the code and then perform the modification. This low cognition level is indicative of the participant reading the code in an attempt to understand what task the code performed. Once this was sufficiently understood, the modification could then be attempted. After that point they began to operate at higher cognitive levels. This pattern was similarly repeated with all participants who did not perform a code inspection prior to modifying task.

Figure 2 graphs participant 5's modification utterances. It shows that participant 5 started with a small number of utterances in the low cognitive levels and then moved into the higher cognitive levels: synthesis, application and evaluation, remaining there for a large portion of the modification time.

Prior to performing the modification task, participant 5 had performed a CBR inspection. Figure 5 graphs participant 5's utterances from the CBR inspection. The graph shows participant 5 started the inspection with a small number of utterances in the lowest cognitive levels and then moved into the higher cognitive levels: Analysis and Evaluation.

Figure 3 shows participant 1 started with very few low-level utterances and then moved into the higher categories within the taxonomy. Before carrying out the code modification, participant 1 had performed an ad hoc inspection on the class. The participant also had a vast number of Uncoded utterances while performing the modification. The recording indicates this participant talked about unrelated things while performing the code changes.

In Figure 4, participant 17's utterances during the modification reflect a very similar story to those already described. Having performed, in this case, an ADR code inspection prior to the modification, the participant started with a small number of low-level cognitive utterances and then moved into the higher cognitive levels.

Figure 6 displays participant 1's utterances from an ad hoc code inspection. The ad hoc inspection technique is without structure and the inspector operates mostly in the lowest cognition levels. Comparing this to participant 15's modification utterances, Figure 1, both start at the very low cognitive levels. The two graphs appear very similar in the way their utterances are spread through the different categories, and yet one is from an inspection, Figure 6, and the other is from a modification, Figure 1. During the inspection, participant 1 is working to understand the code while participant 15 is working to add functionality to the code. As they are attempting to understand the code they operate at similar cognitive levels. One participant is looking for defects, the other looking to add functionality yet the cognitive levels are very similar.

6 Discussion

These results demonstrate that performing a code inspection prior to adding functionality impacts novice developers' cognitive levels. Moreover, the various inspection techniques used affects developers cognitive levels differently.

Figure 6 shows that participant 1 consistently operated in the lower cognitive levels during the inspection. They were attempting to understand what the code does. When comparing it to what the code should have done, they moved into the higher cognition levels.

Figure 1 shows participant 15, who did not perform an inspection prior to modifying the code, largely operated at the lower cognitive levels. This was similar to participant 1's utterances shown in Figure 6. Notably, both participants operated at similar cognitive levels yet were performing very different tasks: participant 1 was performing an ad hoc code inspection and participant 15 was attempting to add functionality. The ad hoc inspection technique, used by participant 1, is unstructured; the inspectors must use their own experience to successfully execute the inspection. The cognitive levels experienced when performing an ad hoc inspection are similar to those experienced when adding functionality to unfamiliar code. In the case of the ad hoc inspection, no direction is given and in the case of the functionality being added, the code is unknown and must first be understood in order for it to be modified. However, when participant 1 moved into the modification task, they operated at the higher cognitive levels, application and synthesis, from very early unlike participant 15 when they were performing th0065 same task (as noted earlier, participant 1 also talked about unrelated things while performing the tasks).

Participant 5, shown in Figure 5, commenced their code inspection with a small number of utterances in the low cognitive levels and then moved into the higher levels. At the higher cognitive levels, as with participant one, they were judging the code for correctness. However, participant five operated more consistently and for a longer time period at these higher cognitive levels than participant 1 did.

Participants using the CBR inspection technique operated at higher cognitive levels and when they moved to add new functionality they continued to operate at the higher cognitive levels: application, synthesis and evaluation more consistently and for longer time periods. The CBR inspection technique facilitated higher cognitive levels within the inspection and the inspector, when making modifications, appeared to continue to function at these higher levels.

We found that in our study of novice developers' cognitive development, during the practical-based skill training exercise, the developers who performed the inspection prior to modification tended to operate at higher cognitive levels from very early on while those who had not performed the inspection tended to operate at lower cognitive levels for longer periods of time.

The results highlight the important role software inspections can play in increasing developer comprehension of a system. They also support the notion that when introduced to a new program or code, one must first go through an initial stage of low cognition levels to gain a basic fundamental understanding of the code, its operands and operations. For the novice developer, the less structured the process for working through this stage, the longer they operate at the lower cognitive levels of the taxonomy. Conversely, the more structured the technique used to familiarise themselves with the code they are working with, the faster they move into the higher cognitive levels of the taxonomy.

The CBR inspection structure facilitates inspectors to function within the highest cognitive levels, above that of both the ad hoc and ADR. This is due to the question and answer nature of the checklist, which requires the inspector to judge the code for correctness.

As software systems continue to simplify the user experience while increasing the complexity for developers, it is important that effective methods are developed to assist novice developers joining these teams to understand the code-bases they are working on as quickly as possible. This will service quality education with skilled graduates that meet the ICT industry needs.

7 Conclusions

This study, and ultimately the three major stakeholders, academia, industry and students, could all benefit from collaborating on future research investigating the benefits of inspections to improve collaborative design cognition and extending team knowledge.

Industry's benefit from collaboration with academia would be in assisting to produce higher quality graduate developers. These graduates' skills would have already been tried and tested in the environment of with industry-based code. This could aid in reducing costs related to graduate training and also reduce the amount of productive time senior developers lose when answering rudimentary questions from new developers.

Collaboration between academia and industry would result in students also benefitting. Prior to moving into the work force, students will have seen and worked on industry based code. Novice developers' exposure to this type of code would create an awareness of the complexities of the code they will be working with when they move from academia into industry. Their education would have covered both the theoretical side and the industry side of issues faced by developers.

The call must be made for increased collaboration between industry and academia. The use of ICT continues to become more and more ubiquitous and the underlying complexities of ICT continue to increase. Collaborative research between academia and industry into effective reading strategies to improve developer comprehension is essential in raising the quality of software being produced by increasing the quality of software development graduates.

The disconnect between academia, their ICT graduates and ICT industries is as common as the gap between business objectives and IT solutions. Despite the strong shortage of ICT skilled professionals in all industries, academia has a hard time creating ICT graduates that meet industry needs. Currently, existing ICT education and the rest of the ICT industry throughout the world are out of sync. For example, the evolution cycle of Technology is 6 months, but in academia, most curriculums change approximately every 3-5 years. Without a collective academic industry learning effort, students will study outdated technology and practices that will be even more outdated in 3 years time, when they graduate. Therefore, the collective effort will help to keep pace between ICT revolutions and state-of-the-art education, enabling global knowledge, networked intelligence, and extended knowledge to penetrate the educational sector.

References

1. Anderson, L.W., Krathwohl, D.R., Airasian, P.W., Cruikshank, K.A., Mayer, R.E., Pintrich, P.R., Raths, J., Wittrock, M.C.: A Taxonomy for Learning, Teaching, and Assessing: A Revision of Bloom's Taxonomy of Educational Objectives. Longman, New York (2001)
2. Basili, V.R.: Evolving and packaging reading technologies. Journal of Systems Software 38(1), 3–12 (1997)
3. Bergantz, D., Hassell, J.: Information relationships in prolog programs: how do programmers comprehend functionality? Int. J. Man-Mach. Stud. 35(3), 313–328 (1991)
4. Bloom, B.: Taxonomy of Educational Objectives Cognitive Domain. David McKay Company, Inc. (1956)
5. Brooks, R.: Towards a theory of the comprehension of computer programs. International Journal of Man–Machine Studies 18(6), 543–554 (1983)
6. Cooper, D., von Konsky, B., Robey, M., McMeekin, D.A.: Obstacles to comprehension in usage based reading. In: Proc. 18th Australian Conference on Software Engineering (ASWEC 2007), pp. 233–244 (2007)

7. Dunsmore, A.: Investigating effective inspection of object-oriented code, PhD thesis, Strathclyde University, U.K (2002)
8. Dunsmore, A., Roper, M., Wood, M.: Systematic object-oriented inspection - an empirical study. In: ICSE 2001: Proceedings of the 23rd International Conference on Software Engineering, pp. 135–144 (2001)
9. Dunsmore, A., Roper, M., Wood, M.: The development and evaluation of three diverse techniques for object-orientated code inspection. IEEE Transactions on Software Engineering 29(8), 677–686 (2003)
10. Ericsson, K.A., Simon, H.A.: Protocol Analysis. The MIT Press, Cambridge (1993)
11. Fagan, M.E.: Design and code inspections to reduce errors in program development. IBM Systems Journal 15(3), 182–211 (1976)
12. Fisher, C.: Advancing the study of programming with computer-aided protocol analysis, pp. 198–216 (1987)
13. Gilb, T., Graham, D.: Software Inspection. Addison–Wesley, Wokingham (1993)
14. Humphrey, W.: A Discipline for Software Engineering. Addison–Wesley, Boston (1995)
15. Kelly, T., Buckley, J.: A context-aware analysis scheme for Bloom's Taxonomy. In: ICPC 2006, Proceedings of 14th IEEE International Conference on Program Comprehension, pp. 275–284 (2006)
16. Laitenberger, O., DeBaud, J.: An encompassing life cycle centric survey of software inspection. Journal of Systems and Software 50(1), 5–31 (2000)
17. Littman, D., Pinto, J., Letovsky, S., Soloway, E.: Mental models and software maintenance. Journal of Systems Software 7(4), 341–355 (1987)
18. von Mayrhauser, A., Vans, A.: Identification of dynamic comprehension processes during large scale maintenance. Transactions on Software Engineering 22(6), 424–437 (1996)
19. McMeekin, D.A., von Konsky, B.R., Chang, E., Cooper, D.J.A.: Checklist Based Reading's Influence on a Developer's Understanding. In: Proc. 19th Australian Conference on Software Engineering (ASWEC 2008), pp. 489–496 (2008)
20. Moore, D.S., McCabe, G.P.: Introduction to the Practice of Statistics, 4th edn. W.H. Freeman, New York (2002)
21. Shull, F., Rus, I., Basili, V.: Improving software inspections by using reading techniques. In: ICSE 2001: Proceedings of the 23rd International Conference on Software Engineering, pp. 726–727 (2001)
22. Siy, H., Votta, L.: Does the modern code inspection have value? In: Proceedings of IEEE International Conference on Software Maintenance, pp. 281–289 (2001)
23. Sjoberg, D.I.K., Hannay, J.E., Hansen, O., Kampenes, V.B., Karahasanovic, A., Liborg, N., Rekdal, A.C.: A Survey of Controlled Experiments in Software Engineering. IEEE Transactions on Software Engineering 31(9), 733–753 (2005)
24. Tyran, C.K., George, J.F.: Improving software inspections with group process support. Communications of the ACM 45(9), 87–92 (2002)
25. Xu, S., Rajlich, V.: Cognitive process during program debugging. In: Proc. Third IEEE International Conference on Cognitive Informatics, pp. 176–182 (2004)
26. Xu, S., Rajlich, V.: Dialog-based protocol: an empirical research method for cognitive activities in software engineering. In: Proc. of International Symposium on Empirical Software Engineering (2005)
27. Xu, S., Rajlich, V., Marcus, A.: An empirical study of programmer learning during incremental software development. In: Proc. Fourth IEEE Conference on Cognitive Informatics (ICCI 2005), pp. 340–349 (2005)

New Possibilities of Intelligent Crisis Management by Large Multimedia Artifacts Prebuffering

Ondrej Krejcar and Jindrich Cernohorsky

VSB Technical University of Ostrava, Centre for Applied Cybernetics,
Department of measurement and control, 17. Listopadu 15,
70833 Ostrava Poruba, Czech Republic
Ondrej.Krejcar@remoteworld.net, Jindrich.Cernohorsky@vsb.cz

Abstract. The ability to let a mobile device determine its location in an indoor environment supports the creation of a new range of mobile information system applications. Our goal is to complement the data networking capabilities of RF wireless LANs with accurate user location and tracking capabilities for user needed data prebuffering. We created a location based system enhancement for locating and tracking users of our control system inside the buildings. User location is used for data prebuffering and pushing information from server to user's PDA. All server data is saved as artifacts (together) with its position information in building. The accessing of prebuffered data on mobile device can highly improve response time needed to view large multimedia data. This fact is very important for new possibilities of intelligent crisis management. Rescuers can handle with new types of artifacts which can increase rescue possibilities.

Keywords: Crisis Management, Prebuffering, Localization, PDPT Framework, Wi-Fi, 802.11b, MDA, Response Time, SQL Server Mobile.

1 Introduction

The usage of various mobile wireless technologies and mobile embedded devices has been increased dramatically every year and would be growing in the following years. This will lead to the rise of new application domains in network-connected PDAs that provide more or less the same functionality as their desktop application equivalents. We believe that an important paradigm is context-awareness. Context is relevant to the mobile user, because in a mobile environment the context is often very dynamic and the user interacts differently with the applications on his mobile device when the context is different. Context-awareness concepts can be found as basic principles in a long-term strategic research for mobile and wireless systems such as formulated in [1]. The majority of context-aware computing to date has been restricted to location-aware computing for mobile applications (location-based services). However, position or location information is a relatively simple form of contextual information. To name a few other indicators of context awareness that make up the parametric context space: identity, spatial information (location, speed), environmental information (temperature), resources that are nearby (accessible devices, hosts), availability of

M. Ulieru, P. Palensky, and R. Doursat (Eds.): IT Revolutions 2008, LNICST 11, pp. 44–59, 2009.

resources (battery, display, network, bandwidth), physiological measurements (blood pressure, heart rate), activity (walking, running), schedules and agenda settings.

We focus mainly on RF wireless networks in our research. Our goal is to complement the data networking capabilities of RF wireless LANs with accurate user location and tracking capabilities for user needed data prebuffering. This property we use as an information ground for extension of information system. The remainder of this paper describes the conceptual and technical details.

2   Basic Concepts and Technologies of User Localization

Among the many location systems proposed in the literature, the most effective are those based on radio location techniques that exploit measurements of physical quantities related to radio signals travelling between the mobile terminal (MT) and a given set of transceivers whose location is known, e.g., base stations (BSs) and/or navigation satellites. Radio signal measurements are typically the received signal strength (RSS), the angle of arrival (AOA), the time of arrival (TOA), and the time difference of arrival (TDOA).

Radio location techniques are generally classified into two categories: modified and unmodified handset solutions. The former techniques require some adjustment to be implemented in existing handsets, while the latter ones only need modification at the BS sites or switching centers. It is apparent that installing a global positioning system (GPS) receiver on each handset seems to be the most straightforward positioning approach. However, additional hardware and required computational burden reduce the power efficiency and increase the weight, size, and cost of the MTs. Moreover, the GPS receiver needs the simultaneous visibility of at least four satellites, which is not guaranteed in indoor and urban environments. To improve reliability and reduce time to position fix, wireless network information can be combined with satellite positioning, as suggested in assisted-GPS (A-GPS) techniques. Nevertheless, the biggest drawback remains the number of modifications required at both the handset and the fixed network infrastructure. Therefore, radio location techniques based on less expensive unmodified terminals represent a promising tradeoff between performance and overall implementation complexity.

2.1   Radio Location Techniques in Wireless Communications Systems

The radiolocation techniques are classified into two main groups on the basis of the number of BSs involved in the estimation process, i.e., one BS or multiple BSs. They can be implemented in any wireless communication system, provided that reliable measurements of the physical quantities related to known signals travelling either from the BSs to the MT (downlink) or from the MT to the BSs (uplink) are available. In the downlink case, location measurements are generally made using a reference signal broadcast by all the BSs with the same power.

RSS Algorithm Requiring More Than One BS
The technique described here call for a minimum number of simultaneously available BSs, which could not be always guaranteed in actual environments wherein the

number of signals received at the MT with a sufficient power level may be lower than that required by the location algorithm. RSS positioning algorithm is based on the measurement of the RSS of a known training sequence sent by the MT to NBS different BSs (NBS 2: 3). If the transmit power is known, the distance between a BS and the MT can be estimated using the received power level and a proper mathematical model for the path los s attenuation law. As a signal strength measurement provides a distance estimate, the MT must lie on a circle centered at the BS. By using at least three BSs to resolve ambiguities, the MT position estimation can be identified via a trilangulation technique at the intersection point of the relevant circles. Power control strategies commonly used in wireless cellular systems may, however, hinder the effectiveness of such a technique.

Single-BS Algorithms

Single-BS solutions offer many advantages over multiple BS ones. The coverage by several BSs (i.e., the hearability) is no longer a problem. Finally, the internetwork signaling requirement is significantly reduced. On the other hand, most of these methods are prone to severe performance degradation in NLOS conditions. The cell identification (Cell-ID) technique (as one example) simply identifies the position of the MT with that of the serving BS. While the idea of the Cell-ID is attractive for its simplicity and low implementation costs, its accuracy is inversely proportional to the cell size and could be not adequate for the FCC requirements and the most demanding location services.

2.2 Data Collection, Localization

A key step of the proposed research methodology is a data collection phase. We record information about the radio signal as a function of a user's location. The signal information is used to construct and validate models for signal propagation. Among other information, the WaveLAN NIC makes the signal strength (SS) available. SS is reported to units of dBm. A signal strength of Watts is equivalent to $10*log10(s/0.001)$ dBm. For example, signal strength of 1 Watt is equivalent to 30 dBm. Each time the broadcast packet is received the WaveLAN driver extracts the SS information from the WaveLAN firmware. Then it makes the information available to user-level applications via system calls. It uses the wlconfig utility, which provides a wrapper around the calls to extract the signal information.

The general principle states that if a WiFi-enabled mobile device is close to such a stationary device – Access Point (AP) it may "ask" the provider's location position by setting up a WiFi connection. If the mobile device knows the position of the stationary device, it also knows that its own position is within a 100-meter range of this location provider. Granularity of location can improve by triangulation of two or several visible WiFi APs. The PDA client will support the application in automatically retrieving location information from nearby location providers, and in interacting with the server. Naturally, this principle can be applied to other wireless technologies. The application (locator) is now implemented in C# using the MS Visual Studio .NET 2005 with .NET compact framework and a special OpenNETCF library enhancement [14].

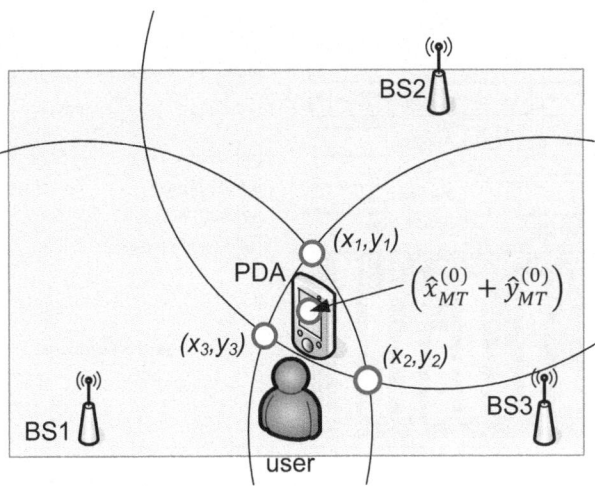

Fig. 1. Localization principle – triangulation

Schema on figure [Fig. 1] describes a localization process. The mobile client gets the WiFi SS of three BSs with some inaccuracy. Circles around the BSs are crossed in red points on figure. The intersection red point (centre of three) is the best computed location of mobile user. The user track is also computed from these measured WiFi intensity level and stored in database for later use.

2.3 Super-Ideal WiFi Signal Strength Equation from Measured Characteristics

The WiFi middleware implements the client's side of location determination mechanism on the Windows Mobile 2005 PocketPC operating system and is part of the PDA client application. The libraries used to manage WiFi middleware are: AccessPoint, AccessPointCollection, Adapter, AdapterCollection, AdapterType, ConnectionStatus, Networking, NetworkType, SignalStrength. Presented libraries are used to manage information about WiFi APs in user PDA nearby. We created special software for scanning user neighborhoods for visible WiFi networks called WiFi Analyzer [Fig. 2].

WiFi Analyzer [13] is a WiFi utility to scan and analyze with PDA for visible WiFi AP. Analyzer show WiFi quality in graph and statistics. Utility allow export measured data to excel. This software allows:

- Display visible WiFi Access Point (802.11b and g standard)
- Analyze visible APs
- Make graph for APs signal measure
- Make statistical data with measured AP signal strength
- Data are updated from 1 to 10 per second
- Insert info about user position to database
- Export option to select data to store in database
- View a graph with signal strength history. (throw Excell)
- Save the measured data to the files on Pocket PC

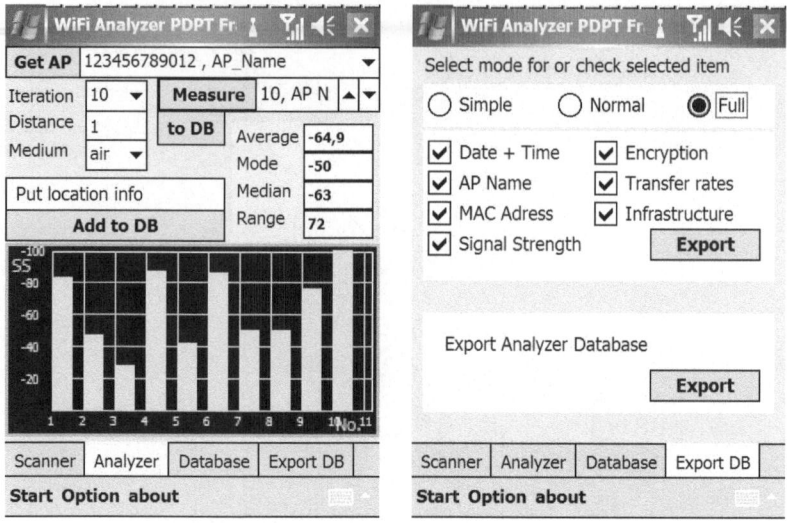

Fig. 2. WiFi Analyzer PDPT Framework – analyzer tab (left) and export DB tab (right)

The WiFi Analyzer software was used to build a database of WiFi APs located in our test environment at campus of Technical University of Ostrava. The campus wireless network is equipped with 32 Cisco AP (December of 2008). Types of these Cisco APs are collect in [Table 1]:

Table 1. Cisco WiFi APs with Transmit Power [mW]

AP Type	*AP* quantities	Transmit Power [mW]			
		50	30	20	10
C1100	24	21	0	2	1
C1130	6	-	6	-	-
C1210	1	1	-	-	-
C1310	1	1	-	-	-

From table summary is evident that 26 APs from 32 APs transmitting with power of 50 mW. We made a simplification to consider only APs with this transmit power. With three APs of them we made a measurements to create a transmit power graph. The SS power was measured by standard PDA device HTC Blueangel to receive a same level of SS as in real case of PDA device usage.

From characteristics at [Fig. 3] is evident the signal strength is present only to 30 meters of distance from base station (this fact is very important for future ideas and real usage because our test environment is not well fitting for use of PDPT Framework).

We used these three characteristics to make a combination of them to get a super-ideal characteristic. Such characteristic need to be compared with ideal theoretic limit

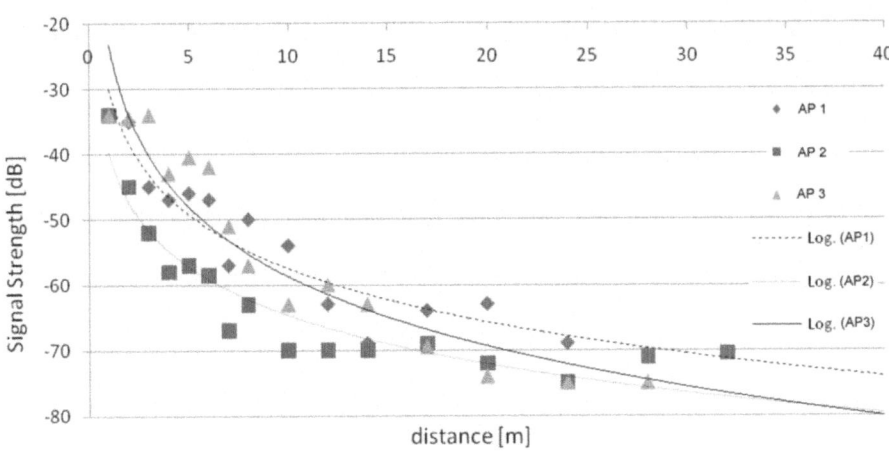

Fig. 3. Signal Strength Graph from three Cisco WiFi APs

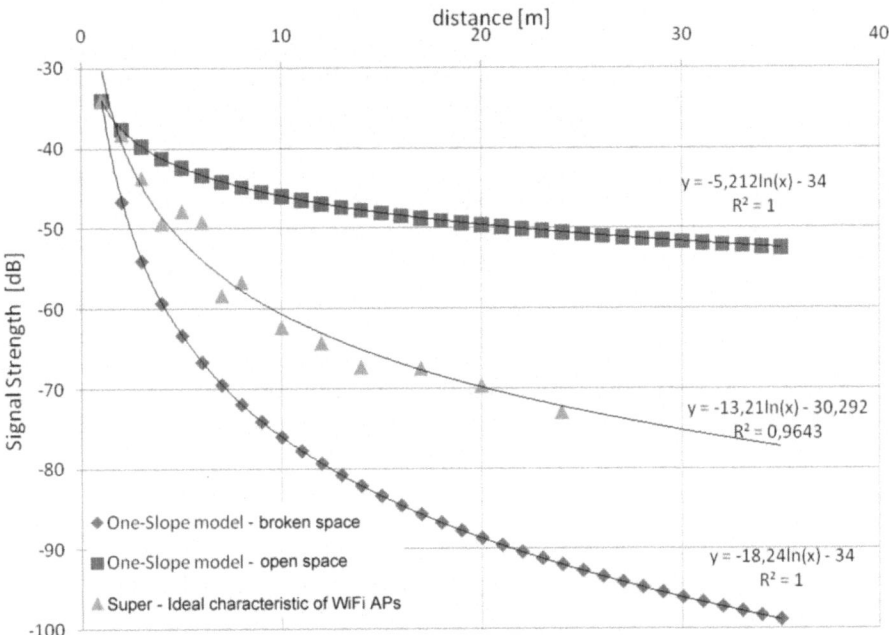

Fig. 4. Real versus Ideal Characteristics of WiFi Signal Strength

characteristics. Equations for computing such characteristics are taken from One-Slope model. See equation (1) for broken space and (2) for open space areas.

$$L(d) = -34 - 42 \, log(d) \text{ for } L_1 = -34 \text{ dB and } n \, 4,2 \qquad (1)$$

$$L(d) = -34 - 12 \, log(d) \text{ for } L_1 = -34 \text{ dB and } n \, 1,2 \qquad (2)$$

Super-Ideal characteristic with two limit theoretic characteristics are in [Fig. 4]. The computed equation for Super-Ideal characteristic is taken as basic equation for PDPT Core to compute the real distance from WiFi SS.

2.4 Predictive Data Push Technology

This part of the project is based on a model of location-aware enhancement, which we have used in created control system. This technique is useful in framework to increase the real dataflow from wireless access point (server side) to PDA (client side). Primary dataflow is enlarged by data prebuffering. These techniques form the basis of predictive data push technology (PDPT). PDPT copies data from information server to clients PDA to be helpful when user comes at desired location. The benefit of PDPT consists of reduction of time needed to display desired information requested by a user command on PDA. Time delay may vary from a few seconds to number of minutes. It depends on two aspects.

First one is the quality of wireless Wi-Fi connection used by client PDA. A theoretic speed of Wi-Fi connection is max 687 kB/s, because of protocol cost on physical layer (app. 30-40 %). However, the test of transfer rate from server to client's PDA, which we have carried out within our Wi-Fi infrastructure provided the result speed only 80 - 160 kB/s (depends on file size and PDA device).

The second aspect is the size of copied data. Current application records just one set of signal strength measurements at the time (by Locator unit in PDPT Client). By this set of values the actual user position is determined by the PDPT server side. PDPT core responds to location change by selection of the artifact to load to PDPT client buffer. The data transfer speed is widely influenced by the size of these artifacts. For larger artifact size the speed is going down.

Theoretical background and tests were needed to determine an average artifact size. First of all the maximum response time of an application (PDPT Client) for user was needed to be specified. A special book [12] of „Usability Engineering" specified the maximum response time for an application to 10 seconds. During this time the user was focused on the application and was willing to wait for an answer. We used this time period (10 second) to calculate the maximum possible data size of a file transferred from server to client (during this period). If transfer speed was from 80 to 160 kB/s the result file size was from 800 to 1600 kB. The next step was an average artifact size definition. We used a sample database of network architecture building plan (Autocad file type), which contained 100 files of average size of 470 kB. The client application can download during the 10 second period from 2 to 3 artifacts. The problem is the time, which is needed for displaying them. In case of Autocad file type we measured this time to average 45 seconds. This time consumption is certainly not acceptable, for this reason we are looking for a better solution. We need to use some basic data format, which can be displayed by PDA natively (BMP, JPG, GIF) without any additional striking time consumption. The solution is in format conversion from any to this native (for PDA devices). In case of sound and video format we also recommend using basic data format (wav, mp3, wmv, mpg).

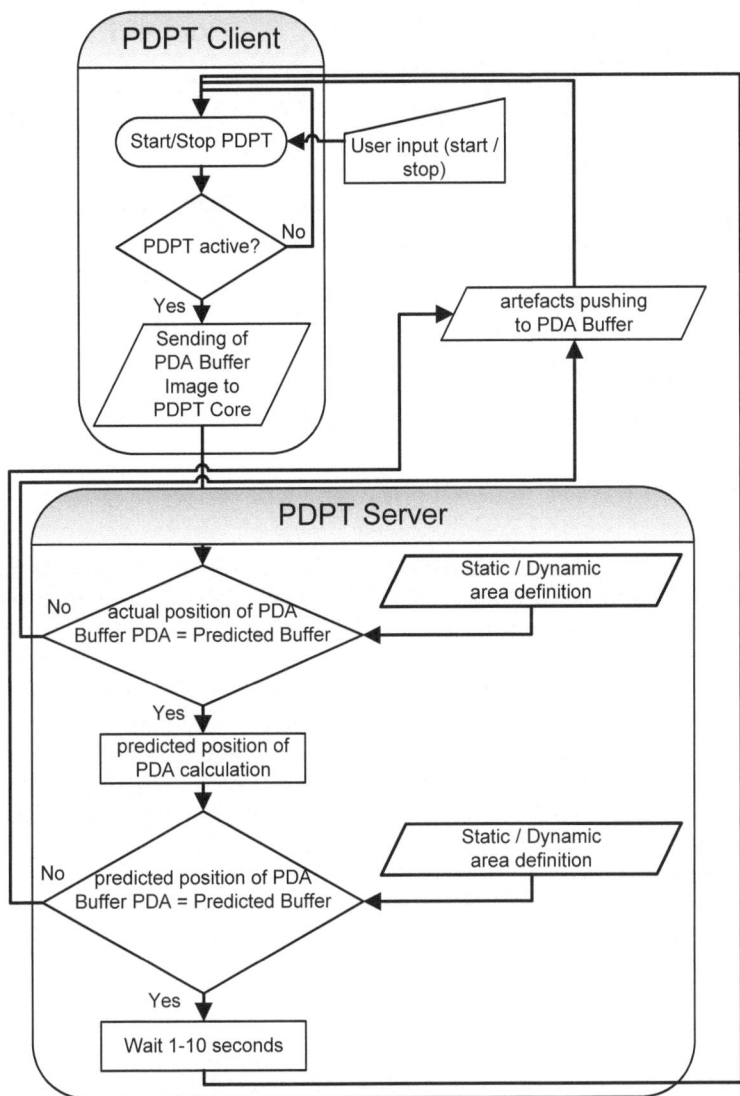

Fig. 5. PDPT Framework – prebuffering architecture

The final result of our real tests and consequential calculations is definition of artifact size to average value of 500 kB. The buffer size may differ from 50 to 100 MB in case of 100 to 200 artifacts.

2.5 Framework Design

PDPT framework design is based on the most commonly used server-client architecture. To process data the server has online connection to the control system. Technology data are continually saved to SQL Server database [2] and [4].

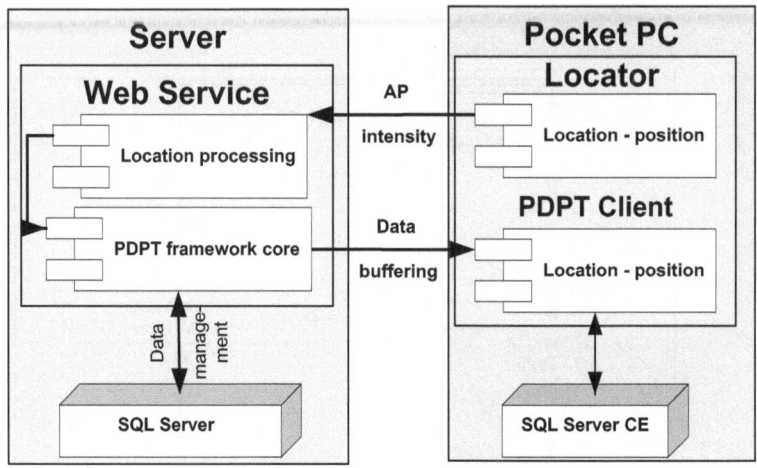

Fig. 6. System architecture – UML design

The part of this database (desired by user location or his demand) is replicated online to client's PDA, where it is visualized on the screen. User PDA has location sensor component, which continuously sends the information about nearby AP's intensity to the framework kernel. The kernel processes this information and makes a decision if or how the part of SQL Server database will be replicated to client's SQL Server CE database. The kernel decisions constitute the most important part of whole framework, because the kernel must continually compute the position of the user and track, and make a prediction of his future movement. After doing this prediction the appropriate data (part of SQL Server database) are pre-buffered to client's database for the future possible requirements. The PDPT framework server is created as Microsoft web service to handle a bridge between SQL Server and PDPT PDA Clients.

2.6 PDPT Client

For testing and tuning of PDPT Core was created the PDPT Client application. This client realizes classical client to the server side and an extension by PDPT and Locator module. Figure [Fig. 7] shows two screenshots from mobile client. The first one (left) show the table of exist artifacts in database on server. The user can select several artifacts by checking and press the "Load To SQL CE" button to push selected artifacts from server to client. This is the classical way how clients can download the data from server. The second one screenshot shows classical view of data presentation from MS SQL CE database to user (in this case the image of Ethernet network in company area plan). Each process running in a PDPT client is measured in millisecond resolution to provide a feedback from real situation. In first case the time window is in lower right side of screen. In second one the upper right side of screen is occupied.

Tabs PDPT and Locator (see figure [Fig. 8]) presents a way to tune the settings of PDPT Framework. In first case (left screenshot) the user must turn on Locator checkbox which mean the measurement of WiFi signals of nearby APs (time of these operations is measured in Locator Time text window). The info about nearby APs is

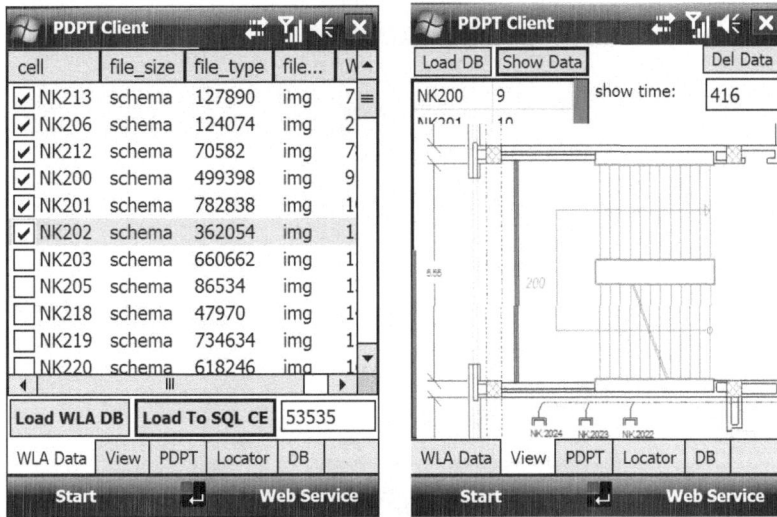

Fig. 7. PDPT Client – Windows Mobile 6.0 application

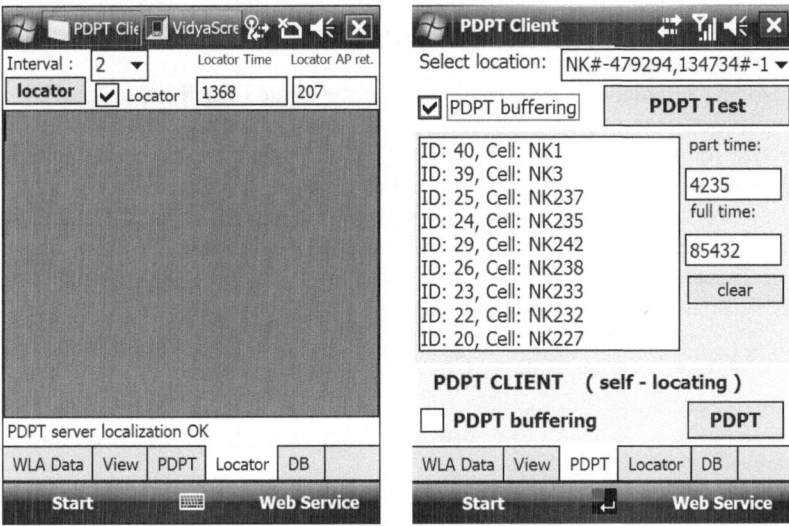

Fig. 8. PDPT Client –Locator and PDPT tabs

send to PDPT server which responds with a number of recognized APs in database (Locator AP ret. Text window). In current case the 7 APs are in user neighborhood, but only 2 APs are recognized by PDPT Server database (info about them is in WLA database). Scanning interval is set to 2 seconds and finally the text "PDPT server localization OK" means the user PDA was localized in environment and is possible to use this position by PDPT Framework core to prebuffer the data to client device. The second case of figure [Fig. 8] is PDPT Client settings tab. The middle section describes the process of prebuffering by logging info. The right side means measure the time of one artifact loading ("part time") and full time of prebuffering.

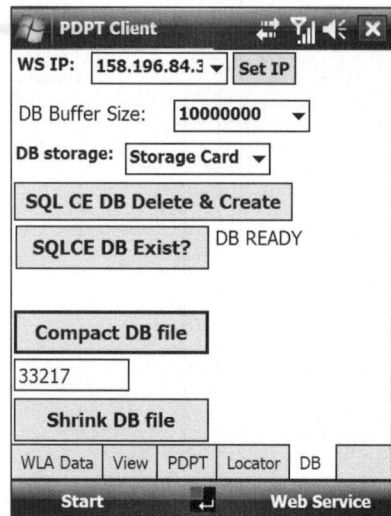

Fig. 9. PDPT Client – Windows Mobile 6.0 application. DB Tab

2.7 PDPT Client – SQL Server CE Database

For managing of database file on PDA device we created small DB manager [Fig. 9].
First *combobox* menu on this tab deal with IP address settings of PDPT server. DB
Buffer size follows on second *combobox*. This size is important for maximum space
taking by prebuffering database on selected data media. Data medium can be selected
on DB Storage *combobox*. For check of database existence the SQLCE DB Exist button
must be pressed. In example the db is ready means the database file exists on selected
location. If such db file does not exist, we need to execute the SQL CE DB Delete &
Create. This buttons can be used for recreating of db file (see program code example).

Example of a C# Program Code – *SQLCE DB Create* button:

```
SqlCeEngine eng = new SqlCeEngine(
                @"Data Source=\SD_CARD\\DbBuffer.sdf");
eng.CreateDatabase();
SqlCeConnection CEcon = new SqlCeConnection(
                @"Data Source=\SD_CARD\\DbBuffer.sdf");
CEcon.Open();
string String = "CREATE TABLE buffer("
+ "Date_Time DateTime not null, "
+ "cell nvarchar(50) not null, "
+ "file_type nvarchar(50) not null, "
+ "file_binary image not null, "
+ "file_description nvarchar(50) not null, "
+ "ID bigint not null "
+ ")";
SqlCeCommand CEcmd = new SqlCeCommand(String, CEcon);
CEcmd.CommandType = CommandType.Text;
CEcmd.ExecuteNonQuery();
```

Compact and *Shrink of DB* file means two options of compacting a database by manual way. The time in millisecond is measured in text box between both buttons. Both of these mechanism are used in prebuffering cycles when the large artifact is deleted from database table, because the standard operation of delete order is not include this technique, so the database file is still has occupied space of deleted artifact. This is due to recover possibilities in Microsoft SQL Server CE databases.

3 User Localization in Intelligent Crisis Management

Many people define crisis management as emergency response or business continuity, while other people will only consider the public relations aspect [3]. They are all partially correct, but true crisis management has many facets. It must be thoroughly integrated into the organization's structure and operations. Achieving an effective level of crisis management requires a thorough internal analysis, strategic thinking and sufficient discussion.

Crisis Management is the umbrella term that encompasses all activities involved when an organization prepares for and responds to a significant critical incident. An effective crisis management program should be consistent with the organization's mission and integrate plans such as Emergency Response, Business Continuity, Crisis Communications, Disaster Recovery, Humanitarian Assistance, etc.

Fireman, police and rescue service are very important part of this crisis management. Management and coordination of this people is now practicable by shortwave communication (radio, transmitter), but new mobile communication technologies as PDA's can level up potential and speed of action in crisis situations. As discussed before, we can locate any people with PDA running client software. But how localization of these people can help them? Advantage is in tracking of these people. For example when fireman arrive to crisis place, his PDA will make an interconnection to crisis management system of building which fireman arrive and the software on PDA will guide the fireman by shortest safe way directly to the centre of problem in the crisis building. In this case function, the PDA act as a navigator and it can help people to make a good orientation around unknown building.

Navigating of rescue people is first but not last possibility which PDA and localization can help. The PDPT Framework can manage large multimedia artifacts as described before. Using of such multimedia files allow to access on PDA for example building plans, pictures of strategic points in environments (electricity case, gas pipe, etc.), video of standard function of machinery or furnishings, etc. These artifacts can be managed to the PDPT Framework by PDPT manager.

3.1 PDPT Framework Data Artifact Management

The PDPT Server SQL database manages the information (artifacts) in the context of their location in building environment. This context information is same as location information about user track. The PDPT core controls data, which are copied from the server to PDA client by context information (position info). Each database artifacts must be saved in database along the position information, to which it belongs.

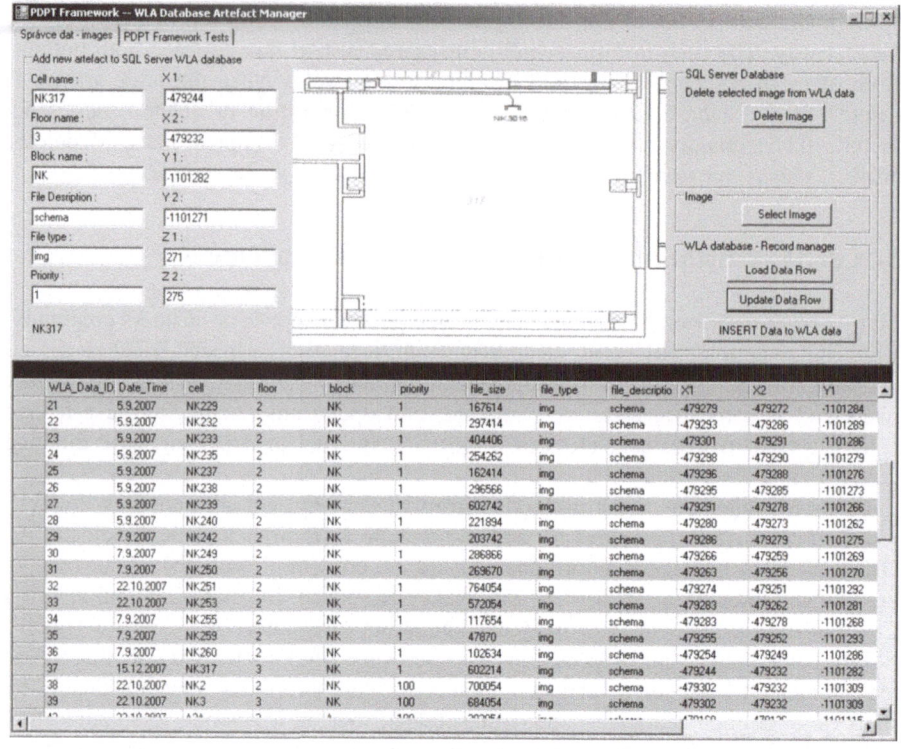

Fig. 10. PDPT Framework data artifact manager

During the creating process of PDPT Framework the new software application called "Data Artifacts Manager" was created. This application manages the artifacts in WLA database (localization oriented database). User can set the priority, location, and other metadata of the artifact. This manager substitutes the online conversion mechanism, which can transform the real online control system data to WLA database data artifacts during the test phase of the project. This manager can be also used in case of offline version of PDPT Framework usage [Fig 10].

The Manager allows to the administrator to create a new artifact from multimedia file (image, video, sound, etc.), and edit or delete the existing artifact. The left side of the screen contains the text field of artifact metadata as a position in 3D space. This position is determined by artifact size (in case of building plan) or binding of artifact to some part of a building in 3D space. The 3D axis is possible to take from building plan by some GIS software like Quantum GIS or by own implementation [4]. The central part represents a multimedia file and right side contains the buttons to create, edit, or delete the artifact. The lower part of the application screen shows actual artifacts in WLA database located on SQL Server.

4   Experiments – Data Transfer Test Using PDPT

We have executed a number of indoor experiments with the PDPT framework using the PDPT PDA application. The main result of utilization of PDPT framework is reduction of data transfer speed. The result of one of these tests is presented here. We focused on the real usage of developed PDPT Framework on wide scale of mobile lightweight devices and its main issue at increased data transfer. For test five mobile devices was selected with different operating system (Microsoft Windows Mobile 5.0, 6.0, 6.1) and a wide scale of memory, display resolution and user interface. For reliability of test we select a test track between two buildings of our university campus environment. User with each device go throw the defined environment where WiFi signal is present (number of AP vary from 1 to 4 visible at a time) ten times for better predicative value. For test we select two main collection of artifact according to their size (medium – 300-500 kB, large – 500-800 kB). During the movement of user the WiFi localization was enabled, so according the user position the artifacts was prebuffered to user PDA buffer (SQL CE database). At destination field we check the number of prebuffered artifacts and this number was compute as Successful rate [Table 2].

Table 2. Data transfer tests description

Test	Type of device	Data artifacts collection	Successful [%]
1	iPAQ h4150	Medium artifacts	84
2		Large artifacts	64
3	HTC Blueangel	Medium artifacts	91
4		Large artifacts	75
5	HTC Roadster	Medium artifacts	92
6		Large artifacts	79
7	HTC Universal	Medium artifacts	96
8		Large artifacts	84
8	HTC Athena	Medium artifacts	98
10		Large artifacts	87

The results surpass our expectations with high quality of successful rate. This rate varies from 84 to 98 % in Athena device case. With large artifacts collection these rates go quite down, but is still very useful for real using of PDPT framework. Is evident the prebuffering techniques can help to use of medium or large artifacts in information systems. If we can transform the real data from information system to artifacts with their positions information, we can improve the transfer rate of used wireless connection and have a better response to users.

5 Conclusions

The main objective of this paper is in the enhancement of control system for locating and tracking of users inside a building. It is possible to locate and track the users with high degree of accuracy. In this paper we have presented the control system framework that uses and handles location information and control system functionality. The indoor location of a mobile user is obtained through an infrastructure of WiFi access points. This mechanism measures the quality of the link of nearby location provider access points to determine actual user position. User location is used in the core of server application of PDPT framework to data prebuffering and pushing information from server to user's PDA. Data prebuffering is the most important technique to reduce time from user request to system response.

We have executed a number of indoor experiments with the PDPT framework using the PDPT PDA Client application. WiFi APs was placed at different locations in building, where the access point cells partly overlap. We have used triangulation principle of AP intensity to obtain a better granularity. Currently, the usability of the PDPT Client is somewhat limited due to the fact that a device has to be continuously powered. If it is not, the WiFi interface and the application cannot execute the location determination algorithm and the PDPT server does not receive location updates from the PDA. This limitation was measured (by battery power level) with results, that is possible to use PDA devices with powered WiFi from 6 to 10 hours without battery replacement. The experiments also show that the location determination mechanism provides a good indication of the actual location of the user in most cases. The median resolution of the system is approximately five meters. Some inaccuracy does not influence the way of how the localization is derived from the WiFi infrastructure. For the PDPT framework application this was not found to be a big limitation for the PDPT framework application as it can be found at chapter Experiments. PDPT framework is currently used in another project of biotelemetrical system for home care named Guardian to make a patient's life safer. [7]. The second implementation of PDPT Framework (Technical University of Ostrava - integrated crisis management of facility management) is currently in development stage.

Acknowledgment. The support for this research work has been provided by the project 102/06/1742: Experimental real-time database testing system, provided by Czech Science Foundation. This work was also supported by the Ministry of Education of the Czech Republic under Project 1M0567.

References

1. Abowd, G., Dey, A., Brown, P., et al.: Towards a better understanding of context and context-awareness. In: Gellersen, H.-W. (ed.) HUC 1999. LNCS, vol. 1707, p. 304. Springer, Heidelberg (1999)
2. Arikan, E.: Microsoft SQL Server interface for mobile devices. In: 4th International Conference on Cybernetics and Information Technologies, Systems and Applications/5th Int Conf. on Computing, Communications and Control Technologies, Orlando, FL, USA, Int. Inst. Informatics & Systemics, July 12-15 (2007)
3. Crisis Management International, http://www.cmiatl.com
4. Horak, J., Orlik, A., Stromsky, J.: Web services for distributed and interoperable hydro-information systems. Hydrology and Earth System Sciences 12(2), 635–644 (2008)
5. Horak, J., Unucka, J., Stromsky, J., Marsik, V., Orlik, A.: TRANSCAT DSS architecture and modelling services. Control and Cybernetics 35, 47–71 (2006)
6. Evennou, F., Marx, F.: Advanced integration of WiFi and inertial navigation systems for indoor mobile positioning. Eurasip Journal on Applied Signal Processing (2006)
7. Janckulik, D., Krejcar, O., Martinovic, J.: Personal Telemetric System – Guardian. In: Biodevices 2008, Insticc Setubal, Funchal, Portugal, pp. 170–173 (2008)

8. Krejcar, O.: Prebuffering as a way to exceed the data transfer speed limits in mobile control systems. In: ICINCO 2008, 5th International Conference on Informatics in Control, Automation and Robotics, pp. 111–114. Insticc Press, Funchal (2008)
9. Krejcar, O.: User Localization for Intelligent Crisis Management. In: 3rd IFIP Conference on Artificial Intelligence Applications and Innovations 2006, AIAI 2006, Athens, Greece, June 7-9, 2006. IFIP, vol. 204, pp. 221–227. Springer, Heidelberg (2006)
10. Krejcar, O.: Benefits of Building Information System with Wireless Connected Mobile Device - PDPT Framework. In: 1st IEEE International Conference on Portable Information Devices, Portable 2007, Orlando, Florida, USA, March 25-29, pp. 251–254 (2007)
11. Krejcar, O.: PDPT framework - Building information system with wireless connected mobile devices. In: ICINCO 2006, 3rd International Conference on Informatics in Control, Automation and Robotics, pp. 162–167. Insticc Press, Setubal (2006)
12. Nielsen, J.: Usability Engineering. Morgan Kaufmann, San Francisco (1994)
13. Olivera, V., Plaza, J., Serrano, O.: WiFi localization methods for autonomous robots. Journal Robotica 24, 455–461 (2006)
14. OpenNETCF - Smart Device Framework, http://www.opennetcf.org
15. Remote World – WiFi Analyzer, http://www.remoteworld.net
16. Salazar, A.: Positioning Bluetooth (R) and Wi-Fi (TM) systems. Journal IEEE transactions on consumer electronics 50, 151–157 (2004)

Paradox in Applications of Semantic Similarity Models in Information Retrieval

Hai Dong, Farookh Khadeer Hussain, and Elizabeth Chang

Digital Ecosystems and Business Intelligence Institute, Curtin University of Technology,
GPO Box U1987 Perth, Western Australia 6845, Australia
{hai.dong,farookh.hussain,elizabeth.chang}@cbs.curtin.edu.au

Abstract. Semantic similarity models are a series of mathematical models for computing semantic similarity values among nodes in a semantic net. In this paper we reveal the paradox in the applications of these semantic similarity models in the field of information retrieval, which is that these models rely on a common prerequisite – the words of a user query must correspond to the nodes of a semantic net. In certain situations, this sort of correspondence can not be carried out, which invalidates the further working of these semantic similarity models. By means of two case studies, we analyze these issues. In addition, we discuss some possible solutions in order to address these issues. Conclusion and future works are drawn in the final section.

Keywords: information retrieval, semantic net, semantic similarity models.

1 Introduction

Semantic similarity models are a series of mathematical models for computing semantic similarity values among nodes in a semantic net [7]. These models are broadly applied in the field of information clustering and retrieval. For their applications in the field of information retrieval, a common characteristic of these models' working procedures can be concluded as follows:

- First of all one or more nodes in a semantic net (normally the component words of the query) are identified by the literal content of a user query.
- Then the semantic similarity values of other nodes in the semantic net to these identified nodes are computed, and those semantically similar nodes are determined and returned based on the values and a threshold.

Thus, the foundation of these theories is built upon the first group of nodes in a semantic net identified by a given user query. However, as a matter of fact, some user queries are ambiguous or over-particular, which do not have corresponding nodes in a semantic net. In other words, the first group of nodes in a semantic net cannot be identified by the user queries. Without the first group of nodes, the semantically similar nodes cannot be determined and returned. As can be seen from the above argument, there is a paradox in these semantic similarity measure models that these

M. Ulieru, P. Palensky, and R. Doursat (Eds.): IT Revolutions 2008, LNICST 11, pp. 60–68, 2009.

modes could be invalid for the ambiguous or over-particular query situations. The objective of this paper is to introduce the paradox in detail.

In the following sections, first of all, we will review the literature with regards to semantic nets and the applications of semantic similarity models in information retrieval. Next, by means of a case study, we will introduce the paradox of these models' applications in information retrieval. Conclusion and future works are drawn in the final section.

2   Related Works

In the section, we briefly review the current literature with respect to semantic nets and semantic similarity models.

2.1   Semantic Nets

A semantic network (net) is a graphic notation for representing knowledge in patterns of interconnected nodes and arcs. It is a directed or undirected graph consisting of vertices, which represent concepts, and edges, which represent semantic relations between the concepts [8].

An example of a semantic network is WordNet©, a lexical database of English. It groups English words into sets of synonyms called synsets, provides short, general definitions, and records the various semantic relations between these synonym sets. Some of the most common semantic relations defined are meronymy (A is part of B, i.e. B has A as a part of itself), holonymy (B is part of A, i.e. A has B as a part of itself), hyponymy (or troponymy) (A is subordinate of B; A is kind of B), hypernymy (A is superordinate of B), synonymy (A denotes the same as B) and antonymy (A denotes the opposite of B).

2.2   Semantic Similarity Models

The semantic similarity models can be categorized into three main types – edge (distance)-based models, node (information content)-based models and hybrid models.

Edge (Distance)-based Models. Edge-based model is based on the shortest path between two nodes in a definitional network, which is a type of hierarchical semantic net in which all nodes are linked with is-a relations. The model is based on the assumption that all nodes are evenly distributed and of similar densities and the distance between any two nodes are equal. It also can be applied to a network structure [6].

The definition is provided by Rada, which is described below:

Let A and B be two concepts represented two nodes a and b, respectively, in an is-a semantic network. A measure of the conceptual distance between a and b is given by

$$\text{Distance}\,(A, B) = \text{minimum number of edges seperating } a \text{ and } b \qquad (1)$$

and the similarity between a and b is given by

$$sim(A, B) = 2 \times Max - Distance(A, B) \qquad (2)$$

where Max is maximum depth of the definitional network.

Leacock et al. [5] consider that the number of edges on the shortest path between two nodes should be normalized by the depth of a taxonomic structure [5], which are

$$Distance\ (A, B) = \frac{minimum\ number\ of\ edges\ seperating\ a\ and\ b}{2 \times Max} \qquad (3)$$

and the similarity between a and b is given by

$$sim(A, B) = -log\ Distance\ (A, B) \qquad (4)$$

The model is based on the assumption that all nodes are evenly distributed and of similar densities and the distance between any two nodes are equal. Additionally, obviously, this model only can be used in a tree-like structure.

For their applications in the field of information retrieval, a document and a query can be represented by two sets of concepts (nodes) respectively in a semantic network. Meanwhile, the query can be transformed into its Disjunctive Norm Form (DNF), which is a group of conjunctive concepts. Thus, the semantic similarity between the document and the query can be measured by computing the distance between the two set of nodes.

The limitations of the edge-based models can be concluded as follows:

- In normal taxonomic or ontological structure, the network density is not regular, which is opposite to the premise of distance-based approach (and);
- The scope of the model only limits in definitional networks, which does not consider some link-types such as part-of, antonyms and so forth.

Node (Information content)-based Model. Information content model is used to judge semantic similarity between concepts in a definitional network, based on measuring their similarity probabilities based on their information content. This model can avoid the defect of edge counting approach which cannot control variable distances in a dense definitional network [7].

The information shared by two concepts can be indicated by the concept which subsumes the two concepts in the taxonomy. Then we define

$$sim(c_1, c_2) = \max_{c \in S(c_1, c_2)} [-\log P(c)] \qquad (5)$$

Where, $S(c_1, c_2)$ is the set of concepts that subsume both c_1 and c_2, and $P(c)$ is the possibility of encountering an instance of concept c.

Similar to the applications of the edge-based models in information retrieval, semantic similarity values between a document and a query also can be converted as the measure of two sets of nodes in a semantic net.

The limitations of the node-based models can be concluded as follows:

- It ignores the information that may be useful (and);
- Many synonyms may have exaggerated content value (and);

- Due to the fact information content values are calculated for synsets as opposed to individual words, it is possible for the information content value to be over-exaggerated in situations, where synsets are made up of a number of commonly occurring ambiguous words

Hybrid Model. Jiang et al. [2] developed a hybrid model that uses node-based theory to enhance the edge-based model, which also considers the factors of local density, node depth and link types [2]. The weight between a child c and its parent concept p can be measured as

$$wt(c, p) = (\beta + (1-\beta)\frac{\overline{E}}{E(p)})(\frac{d(p)+1}{d(p)})^{\alpha}(IC(c) - IC(p))T(c, p) \tag{6}$$

where $d(p)$ is the depth of node p, $E(p)$ is the number of edges in the child links, $\overline{E}$ is the average density of the whole hierarchy, $T(c, p)$ is the factor of link type, and α and β ($\alpha \geq 0$, $0 \leq \beta \leq 1$) are the control parameters of the effect of node density and node depth towards the weight.

The distance between two concepts is defined as follows:

$$\text{Distance}(c_1, c_2) = \sum_{c \in \{path(c_1, c_2) - LS(c_1, c_2)\}} wt(c, p(c)) \tag{7}$$

where $path(c_1, c_2)$ is the set that contains all the nodes in the shortest path from c_1 to c_2, and $LS(c_1, c_2)$ is the lowest concept that subsume both c_1 and c_2.

In some special cases such as only link type is considered as the factor of weight computing ($\alpha=0$, $\beta=1$, and $T(c, p) =1$), the distance algorithm can be simplified as

$$\text{Distance}(c_1, c_2) = IC(c_1) + IC(c_2) - 2 \times sim(c_1, c_2) \tag{8}$$

where $IC(c_1) = -\log P(c)$, and $sim(c_1, c_2) = \max_{c \in LS(c_1, c_2)}[-\log P(c)]$.

Finally, the similarity value between two concepts is measured by converting the semantic distance as follows:

$$\text{Sim}(c_1, c_2) = 1 - \text{Distance}(c_1, c_2) \tag{9}$$

The testing results show that the parameter α and β do not heavily influence the weight computing [2]. The application of this hybrid model in information retrieval is similar to the edge-based models.

It can be observed that the distance computing between two concepts double the information content difference value between their lowest subsumer and the two concepts. However, for instance, for two high level (low information content value) nodes, their lowest subsumer's information content value may be slightly less than or equal to the nodes' values, thus their computed distance is close to zero, and they can be regarded as similar. However, there is a possibility that the two high level concepts could be hugely different. In other words, the computing result could be contradictive to the fact. Therefore, the hybrid model could meet troubles when measuring the similarity between high level nodes, which can be considered as a defect of this model. In addition, as it is the integration of edge-based and node-based models, some defects of these models also appear in the hybrid model, such as exaggerated information content values of synonyms.

3   Case Study for Analysing the Paradox

In this section, we will use two case studies to respectively introduce the invalidity of the semantic similarity models in the scenarios of query ambiguity and over-particularness situation. We choose WordNet© as the semantic net environment.

3.1   Case Study I – Query Word Sense Ambiguity

Nissan® is a well-known Japanese automobile company name, and we want to retrieve the word's meaning by WordNet© in this case study. Once we enter the query term *"Nissan"* into the WordNet© search engine, the search engine can return its glosses and synset relations. In this case, the synset relations include its direct hypernyms, inherited hypernyms and sister terms (Fig. 1). These relations are displayed as a tree-like structure where the *"Nissan"* node is the one of the tree's leaves (because there are no direct hypernyms of the node), its sister terms are the other leaves of the tree which has same joint, and its inherited hypernyms are the branches connecting it to the tree's trunk. Thus, these terms (nodes) and relations construct a semantic net together. In normal cases, the models mentioned above can be used for computing the semantic similarity values between the *"Nissan"* node and the other nodes in the semantic net.

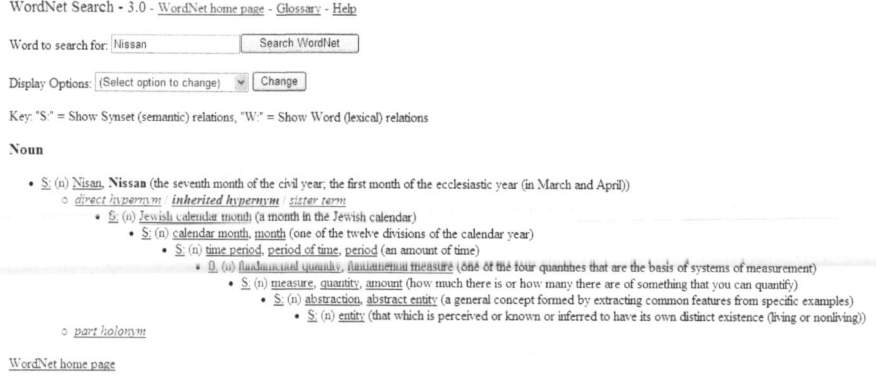

Fig. 1. Retrieved results of *"Nissan"* in WordNet© search engine

However, the returned glosses of *"Nissan"* from WordNet© indicate that it is a religious word (Fig. 1). Obviously this is not the correct meaning of the word that we want to query. Thus, the user-queried *"Nissan"* node cannot be located in the semantic net. Furthermore, those models cannot be applied for finding semantically similar nodes without locating the user-queried node.

The reason why this problem occurs is that there is an ambiguity between the word *"Nissan"* in the field of automobile and in the field of religion. Wordnet© only denotes the religious acceptation of *"Nissan"*, which results in the problem of node mislocation in the semantic net.

In conclusion, this case study illustrates that semantic similarity models cannot be applied for the situation when query words are ambiguous for semantic nets.

3.2 Case Study II – Query Word Over-Particularness

(Westringia) fruiticosa is a sort of Australia's unique plant (Fig. 2). In this case study, we want to query the word "*fruiticosa*" in WordNet©. However, the return result shows that this word cannot be retrieved (Fig. 3). Similarly, the semantic similarity models cannot work in such situation. This is because the word "*fruiticosa*" is so particular that there is no record of this word stored in the WordNet© knowledge- base.

Fig. 2. Interpretation of "fruiticosa" in Wikipedia®

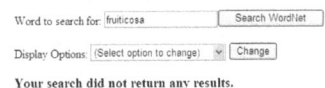

Fig. 3. Retrieved results of "fruiticosa" in WordNet© search engine

Based on the above case studies, we analyse the paradox of semantic similarity models in two primary situations, which are query word ambiguity and over-particularness. These situations contribute to the issue that these models cannot locate the first group of nodes in a semantic net and thus cannot proceed further. There is no methodology provided by the authors of these models to solve this issue. In the next section we will provide some possible solutions to this issue.

4   Possible Solutions

In this section, we will propose possible solutions for the issue of query word sense ambiguity and over-particularness.

4.1 Possible Solutions for Query Word Sense Ambiguity

For the issue of query word sense ambiguity, the most popular approach is to use supervised machine learning approaches, which normally use pre-defined ontologies [4]. There are two possible solutions which can address this issue, as described below:

- One possible solution is to use generic ontologies, which provides common senses of generic words. One significant instance is Cyc knowledge-base, which is a general-purpose repository of common sense concepts and facts [3]. Its application – OpenCyc stores over 47,000 concepts and over 306,000 facts in its knowledge-base (www.opencyc.org).
- Another possible solution is to use domain-specific ontologies, which provide senses for domain-specific terms. One example is Gene Ontology (GO), which uses ontology to annotate gene terms. Gene Ontology database stores over 20,000 terms in the genic field (www.geneontology.org).

These ontologies can provide multiple senses for ambiguous query words in order to reduce the word sense ambiguity. By means of a question-answering module, user can choose the most appropriate sense of a query word, and thus correctly locate the first group of nodes in a semantic net.

However, this methodology has limitations described below:

- Most of these ontologies need to be manually created, which is a labour-intensive and time consuming task and may not necessarily cover the ambiguous words in all domains [1].
- There could be no node in a semantic net that can match the user-selected sense. For instance, in our first case study, although user can choose an appropriate sense of *"Nissan"* – an automobile company's name from an ontology, WordNet© cannot return the relevant result due to the fact that the sense is not stored in its knowledge-base.

4.2 Possible Solutions for Query Word Over-Particularness

For the issue of query word over-particularness, the possible solution could be to use online dictionaries to enrich semantic nets' content. There are two approaches below:

- One possible solution is to use online dictionary APIs to find synonyms for over-particular query words. Then the synonyms are matched with words in semantic nets.
- Another possible solution is to use online dictionary to manually enlarge the glossary volume of semantic nets.

Obviously the first approach is more cost-saving. However, it is found that most online dictionaries do not provide their APIs. Therefore, the first approach may not be feasible during implementation.

5   Conclusions and Further Works

In this paper, we point out and discuss the existing paradox (research issues) in the applications of the semantic similarity models in the field of information retrieval. It is observed that the process of these models involves two common steps: firstly, the nodes which correspond to a user query are located; then the models start to work and compute the semantic similarity values between these identified nodes and other nodes in a semantic net. Thus, locating nodes corresponding to the query word is a pre-requisite of these models. However, as pointed out in this paper, in certain scenarios the process of locating the corresponding nodes to the user query cannot be carried out, and thus the semantic similarity models cannot process further. In order to shed further light on this paradox, we review the three main categories of the semantic similarity models, which are edge (distance)-based models, node (information content)-based models and hybrid models. For each category of these models, we survey their applications and analyze their limitations in the field of information retrieval. Next, we use two case studies to illustrate the issues in detail – *query word sense ambiguity* and *query word over-particularness* which trouble the semantic similarity models. The two case studies are implemented in WordNet© – a typical semantic net environment. In the first case study, we use a query word which can be located but cannot be disambiguated by WordNet©. Due to this reason that the query word's sense cannot be disambiguated, as a result of which, the semantic similarity models could calculate wrong semantic similarity scores. In the second case study, we use a query word which is not stored in the WordNet© knowledge-base. These models cannot work further without node locating.

We discuss several possible solutions for the two issues. For solving the query word sense ambiguity, supervised machine leaning approaches with ontologoies are popular. For solving query word over-particularness, online dictionaries could be used to address this issue. However, by means of detailed analysis, we found that every solution has its own limitations, which cannot ultimately solve the two issues. Therefore, we assert that in order to solve this paradox further and deep research needs to be carried out in the field of semantic similarity models and semantic nets.

References

1. Andreopoulos, B., Alexopoulou, D., Schroeder, M.: Word Sense Disambiguation in Biomedical Ontologies with Term Co-occurrence Analysis and Document Clustering. Int. J. Data Mining and Bioinformatics 2, 193–215 (2008)
2. Jiang, J.J., Conrath, D.W.: Semantic Similarity Based on Corpus Statistics and Lexical Taxonomy. In: International Conference on Research in Computational Linguistics (ROCLING X), Taiwan, pp. 19–33 (1997)
3. Curtis, J.C., Baxter, D.: On the Application of the Cyc Ontology to Word Sense Disambiguation. In: The 19th International Florida Artificial Intelligence Research Society Conference (FLAIRS 2006). AAAI Press, Melbourne Beach (2006)
4. Joshi, M., Pedersen, T., Maclin, R., Pakhomov, S.: Kernel Methods for Word Sense Disambiguation and Acronym Expansion. In: The 21st National Conference on Artificial Intelligence (AAAI 2006). AAAI, Boston (2006)
5. Leacock, C., Chodorow, M.: Combining Local Context and WordNet Similarity for Word Sense Identification. In: WordNet: An Electronic Lexical Database, pp. 265–283. MIT Press, Cambridge (1998)
6. Rada, R., Mili, H., Bicknell, E., Blettner, M.: Development and Application of a Metric on semantic nets. IEEE Transactions on Systems, Man and Cybernetics 19, 17–30 (1989)

7. Resnik, P.: Semantic Similarity in A Taxonomy: An Information-based Measure and Its Application to Problems of Ambiguity in Natural Language. Journal of Artificial Intelligence Research 11, 95–130 (1999)
8. Sowa, J.F.: Semantic Networks. In: Shapiro, S.C. (ed.) Encyclopedia of Artificial Intelligence. Wiley, Chichester (1992)

Physically Based Virtual Surgery Planning and Simulation Tools for Personal Health Care Systems

Firat Dogan and Yasemin Atilgan

Faculty of Engineering, Dogus University, Kadikoy, Istanbul, 34722, Turkey
{fdogan,yatilgan}@dogus.edu.tr

Abstract. The virtual surgery planning and simulation tools have gained a great deal of importance in the last decade in a consequence of increasing capacities at the information technology level. The modern hardware architectures, large scale database systems, grid based computer networks, agile development processes, better 3D visualization and all the other strong aspects of the information technology brings   necessary instruments into almost every desk. The last decade's special software and sophisticated super computer environments are now serving to individual needs inside *"tiny smart boxes"* for reasonable prices. However, resistance to learning new computerized environments, insufficient training and all the other old habits prevents effective utilization of IT resources by the specialists of the health sector. In this paper, all the aspects of the former and current developments in surgery planning and simulation related tools are presented, future directions and expectations are investigated for better electronic health care systems.

Keywords: Virtual surgery physically based modeling, electronic health care, patient specific surgery, simulation tools.

1 Introduction

The computer models of the physical objects get closer to "real" if both visual and behavioral aspects of the models are well defined. Visual reality specifies how model will be understood and viewed by the users, on the other hand behavioral aspects of the model determines physical effects of the interactions or changes by the environmental variables, factors or usually forces. Haptic devices sit between the model and user interaction to let users experience real life experiment. These three important aspects must be contained in a single framework and let users real time interaction in ideal approach.

Virtual surgery is one of the most promising areas which physically based modeling techniques need to be applied carefully. Surgery practices can be supplied by the cadavers or animals to let young surgeons practice enough before any real surgery. A computer simulation system could potentially save time and money and reduce the need for the cadavers and animals and supports ethics in learning surgery. However, surgery simulation must be realistic to be useful tool in surgery with respect to tissue deformation, tools interaction, visual rendering and real time response. For

M. Ulieru, P. Palensky, and R. Doursat (Eds.): IT Revolutions 2008, LNICST 11, pp. 69–78, 2009.

given surgical simulation, soft tissue deformation accuracy and computation time are the two main constraints. Delingette [1] has summarized the different types of applications according to scientific analysis and surgery planning in figure 1.

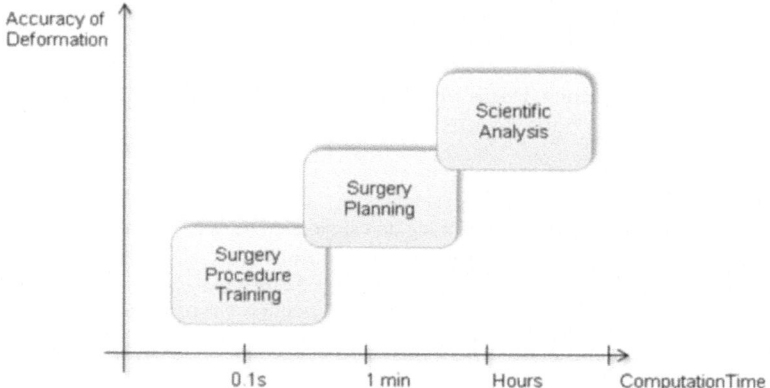

Fig. 1. Required computation time versus accuracy of deformation

Although there are various practices in the area of modeling virtual surgery and deformable objects, physically based modeling methodologies draw attention to achieving desired accuracy. The mass-spring, finite element, mixture of finite difference and boundary element, particle dynamics are the mostly used methods in the area of physically based modeling techniques. In order to supply requirements such as tissue deformation, tool interaction or real time responses, simulation system needs to balance complexities between the each component of the system. In section two, focus will be given to specifically physically based approaches, which are used for modeling virtual surgery. In this category the two widely used classical approaches, mass-spring systems and finite element methods will be presented. In section three current virtual simulators are briefly summarized and future expectations are discussed.

2   Previous Studies in the Area of Physically Based Virtual Surgery Simulators

Physically based methods explicitly use the laws of physics to model objects, calculating internal and external forces in order to determine an object's deformation. Recent research in the context of surgery simulation has developing on physically based models. By modeling the physics as accurately as possible, the aim is to achieve that the models can achieve visual realism for generic models, and perhaps physical accuracy where specific patient data or other object data is known. It should keep in mind that physically based methods uses some assumptions and approximations to model real life object. Material data may not be available in some situations or data captured can be taken in vivo or from cadavers, in some situations data captured from animals and assumptions are made for human organs [2]. Patient specific information is another concern which may change simulation results compared with the real deformation results. Patient's age, sex, health and other factors change simulation parameters and as a result, simulation outputs [3]. Volumetric organ geometry may be extracted from three-dimensional medical images (CT scan images, for instance) of given patients. It may be useful to build a 'standard' model of a given organ, in particular in the context of teaching surgical gestures. On the contrary, when simulator is used for the rehearsing of a given surgical procedure, the anatomy of a given patient must be extracted from his medical images [4].

Several mathematical models have been proposed in the literature for representing the deformable objects and biomechanical behavior of soft biological tissues. The choice of a deformable model must take two contradictory elements: the simulation realism and the computation cost for implementing this model into account.

2.1 Mass-Spring Models

Mass-spring models are easy to construct and both interactive and real-time simulations of mass-spring systems are possible even with today's desktop systems. For applications where the model exhibits non-linear behavior such as soft tissue of cloud dynamics, non-linear or inelastic dynamics can be applied to approximate this type of behavior. Another well-known advantage is their ability to handle both large displacements and large deformations.

Animation of the human face interactions and mimics is the area where mass-spring models are widely used. Terzopoulos and Waters [5] used hexahedral lattice constructed using mass-spring approach to model face and face animation. In this system Terzopoulos and Waters developed 3D hierarchical model of the human face for facial image synthesis and facial image analysis. Terzopoulos and Waters generated facial tissue consisting of three layers of elements representing the cutaneous tissue, subcutaneous tissue and muscle layer for the realistic results. The springs in each layer have different stiffness parameters in accordance with nonhomogeneity of real facial tissue. To account for the incompressibility of the cutaneous fatty tissues, Terzopoulos and Waters included a constraint into each element which minimizes the deviation of the volume of the deformed element from its natural volume at rest. Explicit Euler method solution approach is handled to solve discrete Lagrange equations of motion for the dynamic spring system of coupled differential equations numerically. The forces, accelerations, velocities and positions of each node evaluated in each time step. The Adams-Bashforth-Moulton numerical procedure can be evaluated to maintain convergence of the system and can be used in the case of large stiffness values but its computational complexity per time step that suppresses overall interactive performance of the system. Muscle parameters used to animate physically based model of face extracted directly from video images using physically-based vision techniques and details can be found in reference [6].

Thalmann et al. [7] exploit a generalized mass-spring model of Jansson's et al. which call molecular model where mass points are in spherical mass regions called molecules. Elastic forces are then established between molecules by a spring like connection with integration of properties of real biological materials to define the stiffness of its spring like connections. To validate their model, Thalmann et al. [8] arranged a test setup to compare results with obtained by a Finite Element static analysis. Besides the visual results and displacement of the key points in the modeled object, almost same deformation results are obtained.

3D boundary models of organs are reconstructed from segmented MRI data to model human hip joint as an application of Thalmann et al's molecular spring-mass model. Surface molecules are used to generate contact avoidance forces. An anatomy-based kinematical model of human joints used to simulate motion on the hip joint and evaluate stress and strain on cartilage surfaces. However model used in application is linear elastic and isotropic.

As mentioned before, selection of correct spring constants is a challenging task and is the weakest part of the mass-spring models of soft tissues. In practice, most parameters are mostly determined by trial and error based on the visual results of the simulation. Duysak et al. [9] implemented a learning algorithm using two levels of neural networks to find spring constants for face tissue modeling which produce more realistic and accurate results. Duysak et al's model uses the changes of length and velocity of the springs as input and total spring force as the training signal.

2.2 Finite Element Models

Mass-spring models consider objects as composed of individual particles where as finite element methods (FEM) assumes that material is distributed in a continuum throughout the body. One presents Lagrangian view and the other FEM, considered as an Eulerian view of matter. FEM idea is that continuous function on a domain can be approximated by using other functions defined in smaller domains.

Keeve et al. [10] used displacement-based finite element approach to achieve a result of anatomy-based 3D finite element tissue deformations in implant operations at facial tissues. Generic facial mesh is used to map general anatomical structures to the individual patient data. Using this information, 3D finite element model which is based on a linear formulation is created. The finite element calculations of a 2583 six node prism element took 10.7 minutes using a SGI High Impact workstation.

Niederer and Hutter [11] have taken advantage of reduced integration scheme to decrease finite element calculation of volume integration of the female abdominal cavity deformation. The viscoelastic material with hyperelastic behavior is implemented in deformable organ model.

Vuskovic et al. [12] modeled uterus with fallopian tubes under assumption of homogenous and isotropic materials using FEM formulation. A further simplification is added only concentration on tissue deformation moreover, implementation of cutting or perforation is not included in the proposed model. For mechanical description of soft tissues Vuskovic et al. specified 2nd Piola Kirchoff stress and Green-Lagrange strain tensor relation and used Veronda-Westmann like material definition. For the model based on Veronda-Westmann type material, the constants (material parameters) must be carefully determined. Vuskovic et al. applied pipette tissue aspiration method into pig kidney cortex to measure mentioned constants above.

Inoue et al. [13] implemented surgical simulator system and considered presurgical simulation of liver surgery in which the organ surface and internal vascular system are displayed in desktop environment. Deformation volume rendering, FEM and a haptic device with force sensation feedback are used in the system implementation. In order to stabilize the deformation computation by the FEM with a dynamic solution procedure, the process is managed by distributed processing using shared memory and multiple CPUs.

Kim et al. [14] presented a novel multiresolution approach to balance computational speed and accuracy of computations. When tool-tissue interaction occurs, the "action area" is restricted to only a zone in the vicinity of the tool tip. High resolution model is used governed by user's choice of tool tissue contact location. The rest of the domain is modeled using a relatively coarse model.

3 Discussion of the Physically Based Modeling Approaches in Personal Health Care

Although the continuum mechanics based two common methods, mass-spring and FEM and their applications in virtual surgery are frequently used in literature, one must follow the following standard techniques for fully functional virtual surgery simulator:

- CT, MRI or another type of image accusation of specific organ
- Segmentation of organ images, geometric representation
- Generation of surface meshes
- Volumetric mesh generation
- Calculation of tissue specific material properties
- Implementation of constitutive relations
- Handling force and tools interactions including haptic interfaces
- Rendering of tissues and other geometric objects

Klapan et al. [15] summarized necessary steps and required components of the overall surgery system for computer assisted 3D CAS and tele-3D surgery. The most important question; "execution of personalized features in virtual surgery" arise from particular investigation of steps defined above. A first difficulty appears during segmentation stage of the images obtained from CT or MRI related data. Currently; segmentation of organ geometry handled by operator controlled software where tissue intensity levels are selected by *"try and extract"* method from 2D image data. The main drawback of the user handled segmentation is long operation times where geometric representation may take days up to weeks based on the complexity of the organ or tissue under investigation. The material property of the organ therefore intensity level determines boundaries which are frequently mixed with other surrounding tissues at intermediate points. The relative determination of organ boundaries based on the operator perception or experience is another disadvantage of manual segmentation processes. Campadelli et al. [16] proposed fully automatic gray level based segmentation framework that employs a fast marching technique to overcome difficulties in manual segmentation stage. Segmentation process of female kidney obtained from CT data and meshing stage is presented at figure 2.

The continuum based modeling studies entail element level surface or volumetric meshes to calculate required stiffness matrices that arise from constitutive equations. Mesh generation is another bottleneck in personalized virtual surgery simulations. The quality of the mesh directly affects speed and accuracy of the overall simulation where large numbers of meshes reduce the speed of the simulation or poorly meshed geometry results inaccurate calculations. The segmentation software mostly supports

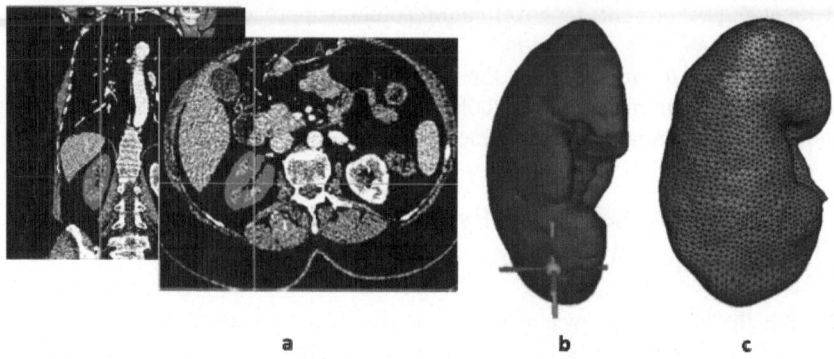

Fig. 2. CT images (a) segmented volumetric object (b) after mesh generation (c)

generation of surface meshes which may require additional software to generate volumetric meshes from surface mesh. Data exchange between different software is required in the cases where volumetric mesh generation is not applicable inside in segmentation software. Analogous to segmentation part, mesh generation is another operator driven process which requires additional afford of medical technician, detailed investigation of overall geometry and mesh corrections are needed at broken or discontinuous parts of the segmented geometric object to produce quality meshes. The mesh reduction and mesh smoothing are two advanced remeshing methodologies which also depend on operator input and experience to reduce total number of elements in calculation domain if real-time simulation experience is desired. Michihiko et al. [17] proposed patient specific modeling method which uses small tetrahedral elements for objective shape and they also introduced "form factor" that indicates degree of complexity of objective shape for automated modeling. Luca et al. applied [18] 3D level set approach to overcome difficulties in semiautomatic procedures at mesh generation stage of blood vessels to introduce patient-specific modeling of blood vessels at clinical level. Reitinger et al. [19] stated that hundred percent error-free automated algorithm is not yet available due to high shape variation, low contrast and pathological data set findings. Information preserving data structure which combines a tetrahedral mesh and binary tree is implemented in their mesh partitioning algorithm.

Human organs, specifically soft tissues have complicated material characteristics such as non linearity in every aspect including contact, geometric and material, anisotropy, large deformation, hysteresis, non homogeneity. Moreover, soft tissues are layered where each layer has its own material properties. Time dependent behaviors of soft tissues are known as viscoelasticity is almost common property for every organ. The selection of correct material property for simulated organ is a very important step which directly affects accuracy of virtual simulator. The measuring techniques of soft tissue material properties can be classified into two main categories: ex vivo and in vivo where most of the measurements are done in ex vivo environment in the past. The organ under investigation is dead in ex vivo measurements. The standard material testing procedures are applicable if material properties are measured outside the body (In vitro measurements). A tissue part generally protected in organ bath or sterilized environment is transferred to the laboratory for measurements. The accurate results can be obtained under known boundary conditions and applied forces with the exact definition of material part in vitro measurements. Miller [20] has designed an unconfined compression experiment to find material constants of the brain tissue. The analytical as well as numerical results to the unconfined compression experiments are presented and validated using commercial FEM software ABAQUS. Abramowitch and Woo [21] implemented uniaxial tension test followed by one hour stress relaxation for the six goat femur-medial collateral ligament to find viscoelastic material properties. The bootstrapping analysis is used in their work to uniquely define material constant. However, recent studies focus on measuring material properties of tissues inside the body (in vivo) specifically in living condition (in situ) due to considerable material property changes of death organs.

The several methods and devices are developed to determine material properties of soft tissues in living condition inside the body. The most frequently used approaches are the aspiration and an indentation experiments in the area of in situ measurements. Kauer [22] applied an aspiration technique to ex-vivo kidney measurements and in-vivo and ex-vivo on human uteri. The in vivo experiments on the human uteri were performed intra-operatively during hysterectomies. Srinivasan and Kim [23] characterized the nonlinear viscoelastic properties of intra-abdominal organs using data from in vivo animal experiments. A total of 10 pigs are used in their indentation experiments where robotic arm controlled stimuli is applied into pigs liver and kidney. Gefen and Margulies [24] presented non-preconditioned and preconditioned relaxation response of porcine brain obtained in vivo, in situ and in vitro at anterior, mid and posterior regions of the cerebral cortex during four mm indentations at 3 and 1 mm/s. They have founded that long term time constant of relaxation significantly decreased from in vivo to in situ modes. Researchers generally construct finite element formulation of soft tissue to compare experimental results with the FEM model and use optimization algorithms to find correct material parameters through inverse solution. The determination of correct material parameters which may vary between different patients is still an open-ended question of personalized virtual surgery applications. Researchers showed that different circumstances including age, gender, life style even psychological conditions change the material properties of living organs. These changes are an important bottleneck of the personalized VR simulators.

All the topics discussed in earlier studies such as surface meshes, volumetric model, material parameters etc. must be combined in a simulation framework where constitutive equations are required for the mathematical presentation of the organ under investigation. The basic blocks of sample simulator framework are given at figure 3. Scientists in the area of virtual surgery simulation have used different constitutive relations to model organ-force interaction and smooth deformation responses. The two frequently used approaches mass-spring and FEM and their applications presented in the previous sections implement different constitutive relations. These relations may either be a simplest linear relation between stress and strain or a strain energy presentation of material such as Ogden, Neo-Hookean, Arruda Boyce or Yeoh type.

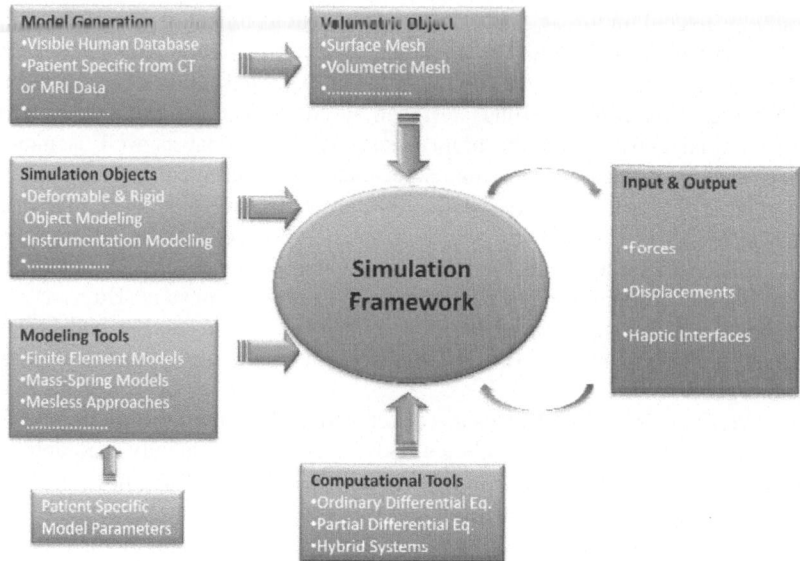

Fig. 3. Parts and steps of the virtual surgery simulator

The fast rendering algorithms and handling user input via haptic interfaces are the other important parts of the virtual surgery simulators. Generally open source rendering and visualization tools such as OpenGL, VTK or Open Inventor type libraries are included in simulation frameworks. Surgeons must feel the interaction forces exerted during tools - tissue interaction as real as possible. Open haptic is one of the open source promising solutions in haptics community for interacting with haptic devices. Real time update rates are possible with the help of computer graphics rendering techniques such as space partitioning, local search and hierarchical data structures. In the point based haptic interaction, only end point of the haptic tool and virtual organ collision is considered. Thin structure of real surgical instruments embarrasses use of point based haptic interactions due to accuracy problems. Tool is modeled using finite line segments in ray based haptic interaction where it has several advantages over point based methods such as feeling torques and rotations.

4   Conclusion and Future Directions

Virtual reality applications have been successfully applied to various areas during the past 50 years. Areas like nuclear energy, aerospace, marinetime, flight simulators which have high risk having direct affect on human life are the main application areas of virtual reality. Even though there has been a growing interest on virtual surgery applications during the past 20 years, the virtual surgery simulators have not still reached the required accuracy. With the use of virtual surgery applications, surgeons may gain the opportunity to test different critical surgery scenarios in a low- cost, ethical environment.

Current simulators use standard geometry data from visible human database. Customization feature is a very important fact for virtual surgery simulators where improvements on automatic segmentation tools will help implementation of personalized data. The ability of transferring volumetric representation of patient specific data into centralized database must be common for future surgery simulator frameworks in a hospital environment.

Material properties of soft tissues change significantly between different patients or depending on various environmental conditions. Furthermore various researchers have measured different material properties for the same body parts even though they have run the same experiments with similar measurement devices. Therefore personalized measurement of material properties is still an open-ended question for virtual surgery simulators.

The simulation results directly affected by mathematical models used in simulator framework. Complexity and solution time eventually increases if new model parameters are added. Depending on the personalized operation procedures; selection option of important tissue properties and required mathematical model within a central framework is one of the feature expected from virtual surgery simulators. Application frameworks like GIPSI, SOFA, PML are still in development to provide required flexibility.

References

1. Delingette, H.: Toward realistic soft-tissue modeling in medical simulation. In: Proceedings of the IEEE, vol. 86, pp. 512–523 (1998)
2. Brouwer, I., Ustin, J., Bentley, L., Sherman, A., Dhruv, N., Tendick, F.: Measuring In Vivo Animal Soft Tissue Properties for Haptic Modeling in Surgical Simulation. Medicine Meets Virtual Reality, 69–74 (2001)
3. Kerdok, A.E., Cotin, S.M., Ottensmeyer, M.P., Galea, A.M., Howe, R.D., Dawson, S.L.: Truth cube: Establishing physical standards for soft tissue simulation. Medical Image Analysis 7, 283–291 (2003)
4. Picinbono, G., Lombardo, J.C., Delingette, H., Ayache, N.: Improving realism of a surgery simulator: linear anisotropic elasticity, complex interactions and force extrapolation. Journal of Visualization and Computer Animation 13, 147–167 (2002)
5. Terzopoulos, D., Waters, K.: Physically-Based Facial Modeling, Analysis and Animation. Journal of Visualization and Computer Animation 1, 73–80 (1990)
6. Magnenat-Thalmann, N., Primeau, E., Thalmann, D.: Abstract muscle action procedures for human face animation. The Visual Computer 3, 290–297 (1988)
7. Thalmann, D., Maciel, A., Boulic, R.: Towards a parameterization method for virtual soft tissues based on properties of biological tissue. Properties of Biological Tissue. In: 5th IFAC 2003 Symposium on Modelling and Control in Biomedical Systems, pp. 235–240. Elsevier Ltd., Melbourne (2003)
8. Thalmann, D., Sarni, S., Maciel, A., Boulic, R.: Evaluation and visualization of stress and strain on soft biological tissues in contact. In: Proceedings of the Shape Modeling International (2004)
9. Duysak, A., Zhang, J.J., Ilankovan, V.: Efficient modeling and simulation of soft tissue deformation using mass-spring systems. In: Proceedings of the 17th International Congress and Exhibition, vol. 1256, pp. 337–342 (2003)

10. Keeve, E., Girod, S., Pfeifle, P., Girod, B.: Anatomy-based facial tissue modeling using the finite element method. In: Proceedings of the 7th conference on Visualization 1996. IEEE Computer Society Press, San Francisco (1996)
11. Niederer, P., Hutter, R.: Fast, Accurate and Robust Finite element Algorithm for the Calculation of Organ Deformations. Virtual Reality based Surgery Simulator Project Report. Swiss Federal Institute of Technology (1999)
12. Vuskovic, V., Kauer, M., Szekely, G., Reidy, M.: Realistic Force Feedback for Virtual Reality Based Diagnostic Surgery Simulators. In: Proceedings of the 2000 IEEE International Conference on Robotics and Automation, pp. 1592–1598 (2000)
13. Inoue, Y., Masutani, Y., Ishii, K., Kumai, N., Kimura, F., Sakuma, I.: Development of surgical simulator with high-quality visualization based on finite-element method and deformable volume rendering. Syst. Comput. 37, 67–76 (2006)
14. Kim, J., De, S., Srinivasan, M.A.: An Integral Equation Based Multiresolution Modeling Scheme for Multimodal Medical Simulations. In: Proceedings of the 11th Symposium on Haptic Interfaces for Virtual Environment and Teleoperator Systems (HAPTICS 2003). IEEE Computer Society, Los Alamitos (2003)
15. Klapan, I., Vranjes, Z., Prgomet, D., Lukinovic, J.: Application of advanced virtual reality and 3D computer assisted technologies in tele-3D-computer assisted surgery in rhinology. Collegium Antropologicum 32, 217–219 (2008)
16. Paola, C., Elena, C., Stella, P.: Fully Automatic Segmentation of Abdominal Organs from CT Images Using Fast Marching Methods. In: Proceedings of the 2008 21st IEEE International Symposium on Computer-Based Medical Systems. IEEE Computer Society, Los Alamitos (2008)
17. Michihiko, K., Masaya, J., Norio, I., Maki, K.: Patient-specific modeling based on the X-ray CT images (new meshing algorithm considering bony shape and density). Japan Society of Mechanical Engineers 72 (2006)
18. Antiga, L., Ene-Iordache, B., Remuzzi, A.: Computational geometry for patient-specific reconstruction and meshing of blood vessels from MR and CT angiography. IEEE Transactions on Medical Imaging 22, 674–684 (2003)
19. Reitinger, B., Bornik, A., Beichel, R., Schmalstieg, D.: Liver surgery planning using virtual reality. IEEE Computer Graphics and Applications 26, 36–47 (2006)
20. Miller, K.: Constitutive model of brain tissue suitable for finite element analysis of surgical procedures. Journal of Biomechanics 32, 531–537 (1999)
21. Abramowitch, S.D., Woo, S.L.Y.: An improved method to analyze the stress relaxation of ligaments following a finite ramp time based on the quasi-linear viscoelastic theory. Journal of Biomechanical Engineering-Transactions of the Asme 126, 92–97 (2004)
22. Kauer, M., Vuskovic, V., Dual, J., Szekely, G., Bajka, M.: Inverse finite element characterization of soft tissues. Elsevier Science Bv, 275–287 (2002)
23. Kim, J., Srinivasan, M.A.: Characterization of viscoelastic soft tissue properties from in vivo animal experiments and inverse FE parameter estimation. In: Duncan, J.S., Gerig, G. (eds.) MICCAI 2005. LNCS, vol. 3750, pp. 599–606. Springer, Heidelberg (2005)
24. Gefen, A., Margulies, S.S.: Are in vivo and in situ brain tissues mechanically similar? Journal of Biomechanics 37, 1339–1352 (2004)

The Primacy of Paradox

John Boardman

A Conference sponsored by ICST
December 17th – 19th 2008
Venice, Italy
www.itrevolutions.org

1 Introduction

Our world today is rife with systems and it's my bet that no amount of revolutions, IT or social, will rid us of them. *Au contraire*, all of our efforts are being directed at bigger, better, smarter systems. Special effort is being directed at a kind of system that makes ready use of a plethora of existing or legacy systems, having them work together in new ways forming what people are calling a Systems of Systems (SoS)[1]. These are new wholes greater than, smarter than, and more potent than not only any of the constituent systems but even the sum of them, however sum is defined.

Systems are ineluctably here to stay. Science however, which gave them to us and allegedly for our benefit, may not be able to lay claim to such permanence. Science, like so many of our present day institutions, systems all, is being rudely and inescapably visited by change, and of a kind that may even sweep science away or minimally make it unrecognizable from that which we know and love today. It is one of my goals in this presentation to say why I believe this and how to prepare accordingly.

2 A Systems Roadmap

Figure 1 shows a roadmap with several notable landmarks. In the top left corner we have systems engineering. This is both a body of scientific knowledge and a rational practice emerging from craft that many argue is the principal delivery vehicle for our contemporary systems. And not merely delivery; systems engineering also take care of maintenance, refreshment, renewal and timely retirement or recycling of these same systems. It is endemic to systems engineering to take care of systems from cradle to grave, or, as we are discovering from the energy-climate era [2], to learn to take care of systems from cradle to cradle.

In the top right corner is the new science, the science of complexity [3], which embraces diverse worlds such as the human brain, colonies of red harvester ants, sexual attraction mechanisms of Papua New Guinea fireflies, weather, climate change, biodiversity, and urbanization.

The remoteness of these two sciences is symbolic of their lack of communion and a spur to those committed to forming, if not a more perfect union of the two, at least a necessary unison of distinct harmonics.

M. Ulieru, P. Palensky, and R. Doursat (Eds.): IT Revolutions 2008, LNICST 11, pp.79–110, 2009.
© ICST Institute for Computer Sciences, Social-Informatics and Telecommunications Engineering 2009

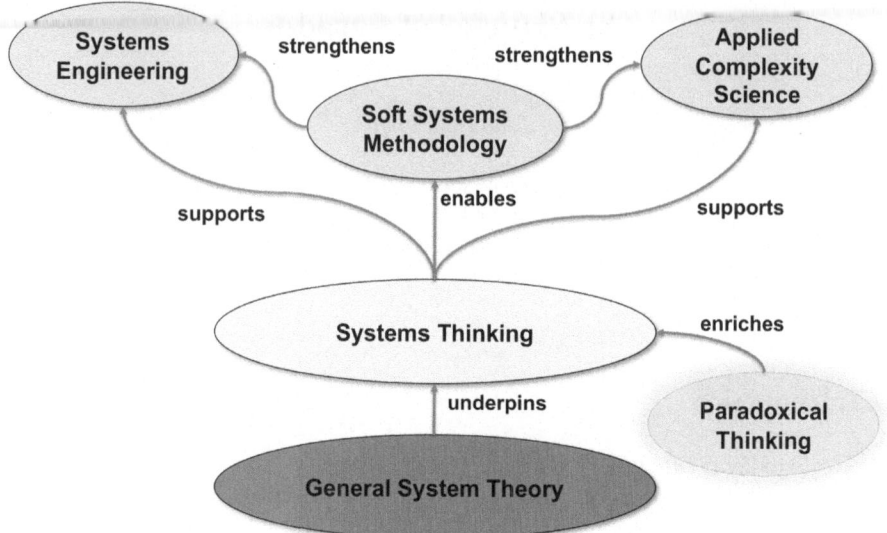

Fig. 1. A Systems Roadmap

Soft Systems Methodology (SSM) occupies the middle ground. This, as I see it, is both a migration of systems engineering, its developer Peter Checkland argued as such [4], and a precursor to the application of complexity science to the most challenging of all systems – ourselves. Perhaps I don't really mean *ourselves* quite as much as I intend to mean *communities of ourselves*, what Checkland termed human activity systems (HAS) and what some are referring to latterly as socio-technological enterprises.

All of these endeavors are underpinned by systems thinking, which is itself underpinned by General Systems Theory.

Thus while each micro-discipline of science, courtiers of the science king or stovepipes within the science park or industrial landscape, support and inform specific scientific endeavors, it is, in my view, systems thinking that provides a specialism in breadth or a horizontal integration mechanism for the universe of systems; a universe that comprises the physical, cyber, intellectual and sociological systems which populate and even crowd our planet and our minds.

A second major goal of my presentation is to validate this underpinning and relate how the horizontality of systems thinking operates and how it might even rescue science from being completely swept away by these irresistible winds of change, though its purposeful dethronement cannot nor should not be avoided.

The final noteworthy piece of this first slide sits unobtrusively to one side. It is paradoxical thinking. Symbolically marginalized, this piece of the landscape is neither innocent nor innocuous. It represents both the winds of change and a source of steering the scientific movement away from dangerous rocks which threaten sinking and drowning. It may even be a sail to direct our intellectual endeavors towards new currents that will take us to fresh discoveries, of knowledge and of ourselves, helping us to create systems better suited to communities of ourselves.

My third and final goal is to describe how paradox is endemic in these systems that heat, flatten and crowd our world, and how paradoxical thinking needs to be incorporated in the systems thinking toolkit or repertoire in order to rescue science from itself and to produce systems better suited to our new world.

3 Thinking about Systems

What can you make of the following assortment: a map of Yellowstone park, an elk, a wolf, a Toyota Prius, and an abstract network? This seemingly arbitrary collection would hardly qualify to be called a system, and yet these elements, and unobvious combinations of them, are precisely what we have chosen to exemplify that term.

If you think about these objects it doesn't take a lot of imagination to come up with a long list of stakeholders with interest in any one or more of them. Here are a few suggestions: geographers, ecologists, biologists, engineers, hunters, campers, metallurgists, and motorists. It might seem obvious at first which stakeholder attaches to which graphic but there are always surprises here.

Stakeholders have an interest in the system; to design it better or to exploit it more to their advantage are two conceivable interests. With that said it is evident that there are at least two kinds of system – the designed, for example the automobile, and the natural, for example Yellowstone.

In recent years there has been some interesting 'leakage' between the two. Designers of systems, having gained new respect for the qualities of design found in natural systems, be this a product of intelligent design or evolution, have sought to produce designed systems that in some way or other tap into these qualities, of resilience, agility or longevity. Designers seek to emulate the inherent, intrinsic attributes of natural systems.

Likewise the designed systems we create increasingly have an impact on nature, with global warming being a prime example of the effects of the so called 'dirty fuels system'. However, on much smaller scales, the increasing population of the automobile and of motorists has impact in terms of road infrastructures, suburbanization, family lifestyles, and so on. All of these changes inevitably leave their fingerprints on nature. Some argue that these effects have removed the invariance of the laws of nature themselves to the presence and interventions of mankind, a premise which has fundamentally underpinned scientific method. If this holds true we are really confronted by a mammoth revolution.

In addition to these points about stakeholders and of the intercourse between natural and designed systems, both of which begin to remove the arbitrariness of these graphics and lead us to see a variety of systems in their own right, I might point out one final and rather crucial point which refers back to the notion of systems of systems (SoS) above.

Both elk and wolf can be found in Aspen; one is prey and the latter is predator. Elk like to feed on young aspen leaves and these disappear rapidly as the elk population increases. Plentiful prey for the wolf ought to reduce the elk population and cause the aspen leaves to flourish. However, scientists have seen more subtle effects than predator becomes prey. In regions of Yellowstone where it appears to elks they have little room for maneuver or probable escape from predatory wolves, these will go untouched by elks and so aspen leaves will shoot up prolifically. It is the fear of the wolf, not merely the presence of wolves that has an effect of Yellowstone's ecology. This is a classic illustration of a system of systems where the relationships between constituent systems – wolves, elks and aspen, have a dynamism that transcends static traditional relationships.

Historically the automobile has been seen as a system, an integrated whole of subsystems, assemblies and parts. There are those who argue that this is a system of systems if the pieces of the system, for example the assemblies, can rightly be termed systems in their own right. But for others this lexicological argument is insufficiently strong. The Toyota Prius however can be truly regarded as an SoS. How so?

This is because the Prius designers recognize and respond to the reality that an automobile, as a system, is playing an adverse role in the wider system that is our planet. By wastefully consuming energy and dumping large quantities of carbon dioxide into the atmosphere the quality of life on the planet is diminishing for all, including the motorist. So these folk have exercised their design skills not only to realize personal transit in the automobile but also energy conservation in that very same system. And they have done this by opening up the functions of the automobile's assemblies to enable each of them to share in a collective responsibility of energy conservation on behalf of the system as a whole. This is turning parts into systems (technically they are called holons) so that these systems now not only provide their autonomous functions, e.g. braking but they have a sense of belonging and connectivity with one another sharing in the system's (i.e. the automobile's) additional responsibility to conserve energy.

Two actions the designers took were firstly to use the energy from braking to generate electrons which could then be stored in the battery and then use that, as opposed to gasoline, for driving as many miles as possible; and, secondly, to convert the kinetic energy realized by the car going downhill to generate more electrons to be stored in the battery for .... well you know. Some call this changing design thinking from focusing on a problem fix i.e. make a car get better gas mileage, to a transformational innovation i.e. make a car produce energy as well as consume it, one consequence of which is better mileage per gallon of gas. Other call this thinking systemically. At some point it will be called a revolution (and the Prius certainly gives many more revolutions per gallon than traditional automobile design!)

The significance of the network icon is to indicate the fundamental difference between a system and an SoS. Whereas the former is patterned conveniently on a hierarchical structure in which even part of a system can be further broken down into sub-parts, the latter is better signified by a network in which there is a greater sense of community at all levels and between levels, and this enables greater belonging of parts and an increased connectivity between them in addition to the sense of autonomy that each does its part but can now execute that role on behalf of the system and whatever revolutionary requirements are placed on that system.

This notion of community is an exciting one. We know it better in application to people systems and perhaps more broadly natural systems rather than the designed ones that people invent. However, it is not always that straightforward to engender serious community among people systems and that brings us to our next section.

4 The Boardman Conceptagon

An important question to ask is "How can systems people, practitioners and academicians, form a community of interest that is mutually supportive and collectively potent as opposed to remaining as a set of stove pipes littering the scientific and industrial landscape?" Put another way, "How can they act like parts of a Toyota Prius rather than the way in which regular parts in a conventional automobile act?" It is true that the latter do a good job, but in our complex world of systems there is more, much more expected of us than was once the case. Because the things that systems people do in their silos affect what others do, and maybe in counter-intuitive and counter-productive ways, it behooves all systems people to use their transcendental talent to work together, as a system, and anticipate the ways in which the systems they produce inter-relate, interact and emerge effects that often counter-point their intended good. My response to this challenge is shown in Figure 2 which is coming to be known as the Boardman Conceptagon.

One of the more important books for systems engineers to read is a work by Harry Goode and Bob Machol entitled, *Systems Engineering*. This is a foundational text, one to which contemporary writers and leading practitioners in this important and rapidly evolving subject area should pay due respect.

A key feature of that work is the distinction made between Interior Design and Exterior Design, a separation that in no way drives a wedge between the two. Rather, systems engineering is concerned not merely with each, but as importantly with the communion between the two.

Thinking abstractly, the words *interior, exterior* and *boundary* (the last being a notional demarcation of the two former) form an essential trio of concepts in the systems engineering lexicon. Likewise, the triple *inputs, outputs* and *transformations*, and another: *wholes, parts* and *relationships* convey ideas that inspire analysis and synthesis of a system's design, both the technical detail within and its suitedness to the system's operating environment.

Not only do these concepts aid the systems engineer in her personal efforts, they represent a bridge to other communities for whom the very same terms also bear much relevance, albeit interpreted vastly differently via specialist make-up.

An abstract thinker might ponder, 'What is the minimal set of triples, covering a sufficient variety and depth of intellectual effort to engage in purposeful systems design and analysis, that would not only support specific technical programs but would also support cooperation and collaboration between several specialisms needed to ensure mission success?'

The reason for asking such a question is twofold: first to form a basis for intelligent debate and effective collaboration between systems people of all walks of life; and, secondly to take a holistic view of the entire mission ensuring that whatever specific pieces the specialists provide, the whole itself is coherent, efficient, and suited for purpose. This collection of ideas is both scale free, covering multiple levels of systems effort, and horizontally integrative, uniting multi-disciplinary labors at any given scale, hence the Boardman Conceptagon.

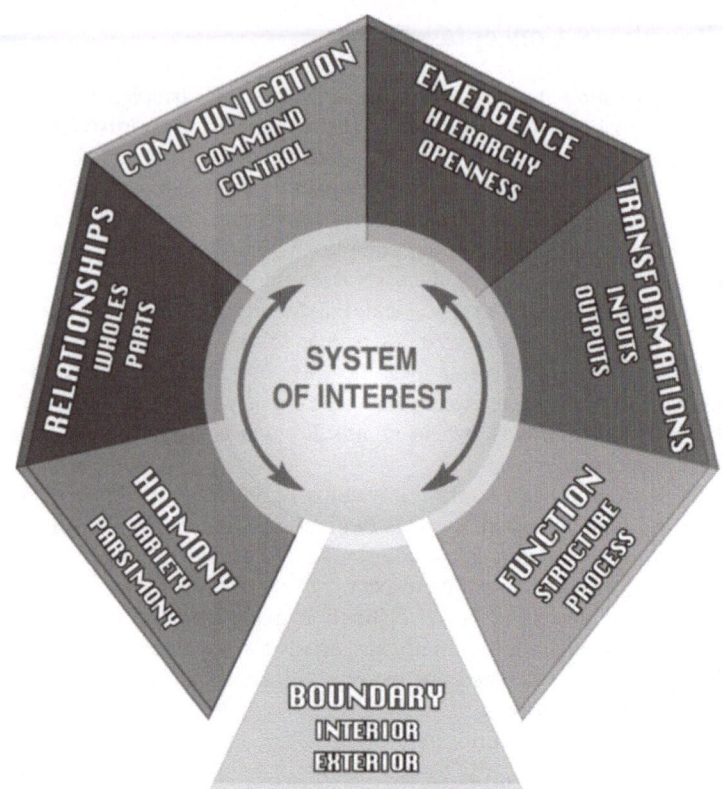

Fig. 2. The Boardman Conceptagon

At the center of this icon is a specific system of interest (SoI). This can be captured initially in a variety of ways: block diagram, flow chart, process map, SysML [5] diagram, or prosaic system description. (Within the systems engineering community and latterly the enterprise architecting community, frameworks exist to guide the overall capture of a system of interest as a collection of models using a variety of modeling paradigms.). Wrapped around this center are seven triples, three of which we have already introduced. The intention is to use each in turn, in any order and with several iterations to explore the system of interest – either analyzing a pre-existing system or synthesizing a new one.

Thus for example consider the triple: boundary, interior, exterior. If we argue that the name of the SoI is the system boundary, e.g. an iPhone, then the inside of the system ought to be obvious, e.g. integrated circuits and the Apple OS X software, or some variant thereof. A crucial element of the exterior would be iTunes with which the iPhone might connect in order to enrich the interior.

One would like to argue that the iPhone was systems engineered. What if systems engineering was the name of the SoI? What then is on the inside? What exactly is systems engineering, and what is on the outside, in the exterior? In other words where does this SoI, systems engineering, prosper and where does it do less well? We will address this issue in the next section.

What if the SoI was a question, e.g. "Who came after Harry Truman?" It would be relatively simple to guide the person being asked this question along a line of reasoning that strongly suggested the pattern to be succession of US Presidents. (Harry Truman was 33rd in the list). Doing so makes it obvious that the answer is Dwight D. Eisenhower, 34th US President. However, a valid response is Doris Day, who was never US President. However, her name comes immediately after that of Harry Truman in Bill Joel's hit record *We didn't start the fire* [6]. Here we see how the interior, the answer to a question in this case, adapts to the exterior, the context of the question. This responsiveness is not an unknown phenomenon to natural systems, and increasingly to designed systems purported to be agile [7].

At first sight it would appear that all the boundary is is a demarcation between the interior and exterior, a concept of little consequence. But then one observes that the interior is what the system is and like all systems it needs to be protected, it needs to be preserved, it needs to prosper. The boundary has a role in all three functions. The exterior may be hostile or docile; it may contain elements which would injure the systems, especially if they got into the interior, or it may contain elements which would be good for the system. Likewise the system may produce waste that may need to be ejected, or recycled, possibly by some other system. The point is that the boundary must permit the ingress and egress of elements in both interior and exterior. And it must do so intelligently. To all intents and purposes this makes the boundary of a system a part of the system (and a part of the exterior) and a system in its own right even, perhaps belonging to both.

This leads to a boundary paradox, our first hint at the paradoxical thinking landmark on our systems roadmap, which can be stated as follows: "You must have a boundary (to nurture and develop specialization in functional expertise, for example). BUT you must also NOT have a boundary (to allow that specialization to be rendered as a service [otherwise why have it] and to allow that expertise to be resourced via interactions with others). So it is possible that a good boundary must both exist and not exist simultaneously. The boundary must *keep* things out and in, but it may also have to *let* things out and in.

This triple and others like it begin to make a line of thinking more formally accessible by their mere existence. Just think what could happen if systems professionals from different walks of life shared their knowledge via this medium! Who knows what new ideas would be generated, possibly even of a transformational nature.

Let us turn to a 2nd triple: command, control and communication – one that is very familiar to the defense community at large and the military services in particular.

When Thomas Stallkamp, former VP of Chrysler, asked his line managers, "How many people and how many firms do you think are involved in the end-to-end process of conceiving, making and selling a Jeep Grand Cherokee?" it took them a while to find the numbers. The answer: 100,000 firms and 2 million people from initial product concept through to a satisfied customer driving her new purchase off a dealer's lot [8]. This answer is startling to most people who have never been involved with the entire life cycle of a product.

But once he recovered from the shock of that answer, Stallkamp posed a telling follow-on question, "Who's managing this enterprise?" The real answer is a paradox – no one is and lots of people are, our second hint at the paradoxical thinking landmark on our systems roadmap.

We must accept the fact that no one sits atop the Chrysler Jeep Grand Cherokee "experience." Perhaps in a titular sense someone does, but in no way can they be said to be its manager. Whoever occupies that office, they are neither a Washington nor a Napoleon commanding thousands and controlling affairs according to their grand strategy. Littered throughout the management hierarchy, or network if you prefer, are hundreds of people, each with their individual spans of influence and concern. But in what ways can this diverse collective be said to be in control of the whole experience when it is probably the case that they are largely unknown to one another? Do these managers perform like ants, somehow supporting excellent behavior for the Cherokee colony? Can control be just as effective, if not more so, if it is distributed rather than precisely located in a central commander? Can we really trust distributing control to a constituency that is largely unaware of the affairs and actions of its neighbors and upon whom a mighty burden of communication must fall in order to rescue order from the grip of chaos?

It is critical to understand whether when we ask "can we really trust distributing control to …?" we are even asking a useful question. When Stallkamp asked who (in particular rather than plurality) is in control of this vast extended enterprise, he was doing what any experienced executive or manager would have done—attempting to establish organizational accountability. A different question, however, may have proved both more illuminating and more useful. The questions we ask are largely a function of our mindset, and that the purpose and benefit of paradox is to confront that mindset—to make us challenge our reliance upon the "conventional" questions and position us to change our mindsets. In so doing, we can release the tension revealed by the paradox and move to the kind of breakthrough thinking that can offer solutions to problems long thought intractable.

Control is to engineers (and managers) what power is to politicians—the ability to influence actions, whether in hard systems or soft ones. Tennyson's The Trinity of Excellence in Leadership, cited by Emile Brolick [9], was neither the first nor the last work to observe that, paradoxically, effective control of others begins with control of self. "Order givers" must not abuse their authority lest they lose it, while "order takers" must exercise self-control when responding—the whole system relies on it.

Current theory includes two poles: "command and control," by which authority "at the top" issues directives that get resolved into executive action by a large group of people; and the "self-organizing" notion of an idea that infects (from the bottom), propagates and galvanizes a group of people who then take action as though they were a unit and had been commanded by a governing authority [10]. These polarities lead to a control paradox: "You must have command and control (to ensure orderliness and conformity to strategic direction). BUT you have must also NOT have command and control (and instead have ground-zero intelligence to foster innovation, tactical opportunism and preservation of self-awareness)."

Put differently:

- authority must exist at the top (for order), but it must also exist at the bottom (for autonomy);
- command must exist (and orders from an external source must be obeyed), but so also must the power to be insubordinate operating alongside a self-will that knows its own order and orders;

- control must operate within a framework that grants liberty to its constituents, but control must also be manifest in the self—as self-control and self-discipline to make the framework work.

The trick for both leaders and managers is to figure out how to embrace and resolve this paradox. Horatio Lord Nelson understood and leveraged it well [11]. He exercised and gave his subordinate commanders:

- **Ordered liberty and Creative disobedience.** To ensure he would be able to seize the initiative in battle, Nelson gave his subordinates (and himself) a clear articulation of his intent and the freedom (within that intent) to deal with situations as they came up, even if it meant ignoring orders.
- **Reciprocal loyalty.** One must give loyalty down the command hierarchy to gain true loyalty (vice obedience through fear). Nelson cared about the welfare of his men, they knew it and reciprocated.

5   Soft Systems Methodology (SSM)

When we introduced the triple of *interior, exterior* and *boundary* we coupled this with the question, "What if systems engineering was the name of the SoI? What then is on the inside? What exactly is systems engineering, and what is on the outside, in the exterior? In other words where does this SoI, systems engineering, prosper and where does it do less well?" We said at that time that we would address this issue in the next section; well, here we are!

Peter Checkland was invited to join the Department of Systems at the University of Lancaster by his close friend Gwilym Jenkins, a noted control systems engineer. The intention was for Peter to do scholarly work that would extend the reach of systems engineering to problems that lay beyond traditional boundaries, more in the socio-technical areas in which culture, politics, and the messiness of humans left its mark.

In retrospect this was a bridge too far, but in fairness to Peter Checkland and in keeping with his strongly scientific approach, rooted in inorganic chemistry and honed in the research laboratories of ICI he began his studies with the codified versions of systems engineering at that time and attempted some form of migration of their essence to a form that better suited the aforementioned messiness of what he later called *human activity systems* (HAS). What finally emerged was Soft Systems Methodology (SSM) and after more than 30 years of exposure to the pernicious currents and howling gales of the messy realities of HAS, which Gwilym Jenkins sought to tame, it has proved rather durable if not the answer that all would wish. Evidently there is room for further development of technique, as one might expect. But this is an excellent opportunity to take stock of Peter Checkland's work, in terms of the essence of its interior and the nature of the exterior to which it has been exposed.

Starting with the latter, this has variously been described by terms such as 'wicked problems [12]', multiple simultaneously tenable viewpoints (STV), ill-structured problems, counter-intuitive behavior [13], and problems that require problem definition. What this means is that the environment has many stakeholders each with valid perspectives on the problem but none in agreement as to what this is precisely. It also means that sincere remedies thoughtfully formulated often have a counter-productive effect since the problems, however defined, have a habit of either ignoring interventions intended to eliminate them or worse responding adversely producing a worsening situation.

Clearly, this is not a happy situation for the would be problem solver, now confronted by a serious dilemma; she is damned if she does nothing and damned when she does. It does little good, though it offers a degree of relief, to read rather humorous accounts of how systems, of both the solution kind and the problem kind (it is often helpful to regard a problem as a system), have a will of their own![14]

Given the fact that SSM has endured, in the face of such perversity, it must at least be intriguing to know what exactly is its essence, what is on the interior of this SSM system, to have given it such fortitude and reasonable popularity. I am happy to give my personal account of this fascinating interior.

The most important feature of SSM is the divide that Checkland promotes between the real world and the systems thinking world. It is as if the only chance of success lies in keeping these two separate. Of course were this separation to be an unbridgeable divide separated by a bottomless abyss which none dare cross, it would add little value. However, bridges are the secret of Checkland's methodology. They are crossed with care and due circumspection as befits the messiness of the real world and the sanctity of the systems thinking world. They are respected by citizens of each world.

Two things never cross this bridge: problems and solutions. This is vitally important, and gives us the 2nd feature of SSM: that the people working away in the ST world are NOT problem solvers; it also means that people battling away in the real world are THE problem owners. They own the problem and the problem remains with them. If the problem is ever solved it is solved in that real world and curiously enough not by means of a solution but rather by means of the problem owners' executive action - which can intelligently be nothing! – that treats whatever problem has been defined arrived at by whatever enlightenment appears, hopefully as a consequence of what does cross the northbound bridge.

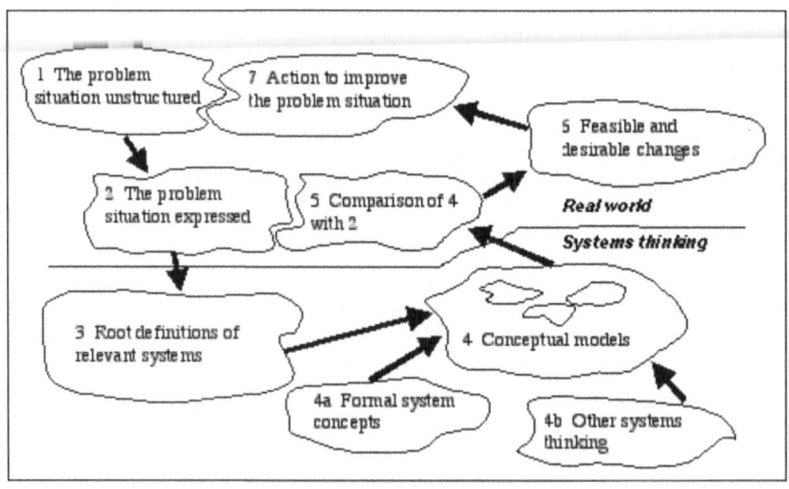

Fig. 3. Checkland's Soft Systems Methodology

If the problem stays permanently as it were in the real world, what is it that crosses the southbound bridge of which systems thinkers can make some sense and give something back, in the form of enlightenment hopefully, to the problem owners? This is a good question. An inspection of Figure 3 might suggest that what heads south is some form of expression of the problem situation, and that is true. But it is only part of the truth. Because in order for the systems thinkers to be effective they must be part of the real world and yet remain aloof from it! They must be accepted in the real world, as legal citizens, when in fact they are more accurately legal permanent residentaliens who never naturalize!

What systems thinkers take with them into their hallowed sanctuary is as deep an understanding of: the problem situation; the problem owners themselves; and, the very culture in which the problem and owners are immersed. Yet the systems thinkers must not be drowned by this understanding; it must not overwhelm them nor bias them in the way that it inevitably does the problem owners. Being part and yet not part of the real world for the systems thinkers is a paradoxical property of the SSM which gives it vitality.

Therefore what accompanies the problem situation expressed on its journey south is this deep understanding held by the systems thinkers and practiced by them in their own world. This leads naturally enough to another question, "What is this practice?". The answer to this provides our 3rd feature of SSM: that the models which the systems thinkers come up with are of IDEALITY NOT REALITY.

It is as if Checkland gave up on building models of reality, the stock in trade of systems engineers, and if this was the case, then it was surely based on this line of thinking. Reality is very messy and any abstraction of it is never going to be informative as to its true nature. This messy reality is subject to enormous subjective interpretation by various stakeholders so even an abstraction of reality based on one perspective will be unrecognizable to a stakeholder with a widely differing perspective. Finally, building models of reality simply does not help since they can in no way advise you what to do – the reality is, you've guessed it, far too messy! Checkland's response was to posit an unreachable ideality which would at least capture, in pure form, a world-view that was indeed idealistic but whose virtue would be to draw people towards the realization of that ideal. Wherever this brought them would still be real but ostensibly far better than the conditions they faced which was giving serious cause for concern.

The technical details of the SSM Practitioners need not concern us at this point, save to say that the effort is twofold: to build, via the problem situation expressions, those ideal systems whose very being would seem to directly follow and whose eventual realization would give rise to an ideal situation. From these statements of systems being follows a practice to define what these systems need to do in order to be. The former effort, leading to root definitions, is strictly definitional with emphasis on nouns, e.g. agents, agencies, and artifacts; the latter effort, leading to conceptual models, is operational, requiring emphasis on verbs. E.g. transactions and transformations.

It is this collection of definitions, models and renewed understanding that heads back north, giving the systems thinkers (aka analysts) an opportunity to play a full part in the real world via their dialogue with the problem owners. Once again there is a key pointer to bear in mind as the dialogue unfolds among analysts and stakeholders, a dialogue enriched by the conceptual models. This refers to the manner of making comparisons between the original problem expression and the conceptual models, and Checkland suggests four distinct ways.

First, the models can be used to suggest a line of ordered questioning. Quite possibly what takes place in the model in no way resembles what presently transpires – but that can be a good thing, and it stimulates fresh thinking by the stakeholders. A second way is to reconstruct the past and compare history with what would have happened had the conceptual models been followed faithfully. In both these cases, the models themselves can be hidden from the stakeholders, so that they are not seized upon as either the answer to their problems or absurd notions that can be flatly rejected bringing an intransigence among stakeholders and a resignation to cope with what they have. A third way is to reveal the models and accompany their presentation with questions about how they differ so much from present reality and why. Finally, an approach known as 'model overlay' can be tried. Here a new set of conceptual models are created this time they are based on reality and are designed to capture as much as possible the way things are. The only rule is that so far as possible they should have the same form as the 'divorced' conceptual models. What this overlay approach does is to highlight the distinctions, which of course is the source of discussion for change. What's more these new models can be reverse engineered into root definitions, and then that can be compared with the one that was obtained from phase 1.

The purpose of this dialogue phase is to generate debate about possible changes which might be made within the perceived problem situation. In practice the work done so far can itself become the subject of debate and change! This requires humility on the part of the analyst: how can he expect stakeholders to change when she herself won't? Changes can be to any of the artifacts of the process, or to the process itself. (But then thinks can only get better!)

Checkland suggests there are three types of change: in structure, in procedures, and in attitude. Structural changes may occur to organizational grouping, reporting structures, or functional responsibility. Procedural changes are to the manner of getting things done, for example the periodicity or medium for reporting. Attitudinal change is of the mind and the heart, and usually less easy to accomplish than the former. The criteria for suggesting and effecting change must be whether they are systemically desirable and culturally feasible. The former refers to a respect for the integrity of the SSM and of all the artifacts this generated; in other words change ought not to be arbitrarily effected simply because 'something has been done' but that something has not been acknowledged. The later criterion shows respect for the problematique and for the stakeholders themselves. Even when change is obvious and agreed upon, it may not actually be deemed implementable simply because of 'conditions'. That is political (and economic) reality. The analyst hasn't necessarily failed; after all he is no longer dealing with buttons in the engineering sense, but belly buttons.

6 Paradoxes of Complexity

We come now to the top right hand corner of our systems roadmap, to the science of complexity. Our interest here has little to do with admiring or recounting the enormous and remarkable achievements in this space over the past half decade or so, summarized sublimely and with unique elegance and authority by James Gleick [15].

It has more to do with revealing the inherent paradoxes that complexity science holds. In this way we can begin to see how the terms *system, complex* and *paradox* form an amazingly powerful trinity.

Our summary of these paradoxes is as follows:

- Complexity is simpler than it first appears
- Simple things exhibit complex behavior
- Little things mean a lot
- Large worlds are smaller than we think
- Significant things are vital and obscure
- Weak relations bring strength and security
- The rich get richer and the poor
- A complex is one and many simultaneously

What we will do is to take a cursory glance at our complex world, with its 6 billion plus people, its intricate food webs, the behavior of some of its less obvious animals – fireflies of Papua, New Guinea, and our own human brains, among many other of life's complexities and in so doing illustrate the paradoxical summaries we cite above. In so doing we hope to emphasize the primacy of paradox as a way of life, our way of life, and set a new direction for scientific thinking, a way that not only benefits from complexity science but also from systems thinking, one that has paradoxical thinking in its core.

6.1 A Weakness Stronger Than Strength

How many people do you know? Is it 10, 50, 100, or more? Some of us know lots of people, others very few. In reality it is harder to really know many people very well. So some of our relationships, relatively few, are deep and regularly maintained. Others are more on the acquaintance level and it is often difficult or too bothersome to keep these latter relations going. Over time they usually whither and die. Their demise at least can be considered at the expense of strengthening the few that really matter, however those are decided.

Some of your 'circle' will know people you don't know, which in a way extends your network. But most of the people you know will know one another. These circles are probably better referred to as a cluster, a reasonably tight knitting together of a close group of friends. Clusters make up a world but they do nothing to make it small. Strong ties hold a cluster together but it is the weak ties that turn a collection of clusters into a small world [16].

It is this paradox, that weak ties are what gives a small world its strength, that for us typifies complexity and systems thinking. It is one of many, as we shall discuss. But lingering with this a little is worthwhile for we must continue to break out of a mindset that insists a network is either a circle of friends or a circuit of transformers, power lines and switchgear. Weak ties explain the computing power of our human brains and the synchronicity of fireflies in a tropical rain forest in Papua, New Guinea. From friends and fuses to fireflies and firing neurons. We are beginning to explore the groundbreaking science of networks, and with it immense opportunities for complexity understanding and systems thinking.

The articulation of a small world architecture using the mathematics of graph theory is comprehensively captured by Duncan Watts [17]. A paper by Watts and his thesis advisor Steve Strogatz, largely free of mathematics, set ablaze huge interest in the phenomenon of small-worldness with its architectural fingerprints being found in diverse fields as ecosystems, natural language and the World Wide Web [18].

At the outset of their work together they sought to introduce random links between a fully ordered network of clusters. Suppose an initial circle (of friends say) has 1,000 dots, each connected to its 10 nearest neighbors. This gives in all about 5,000 links. To this let's add 10 links at random (0.2%). The network is still essentially ordered but is now lightly splashed with randomness. More and more links can be added gradually at random and the effect on clustering and on any perceived small worldness calculated. This is a form of network evolution with strict order being updated with random rewiring. They found that whilst small disturbances had no noticeable effect on clustering it had a devastating effect on small worldness. Initially the degrees of separation was 50 but with a few random links it plummeted to 7. Skeptical of their findings their experimentation continued with greater scrutiny. No matter what they did however they always found that the lightest dusting of the ordered network with random links produced a small world.

In a planet of 6 billion, with each person linked to their 50 nearest neighbors, the number of degrees of separation is of the order of 60 million. Throw in a few random links, a fraction of the total being 0.02 % and the degree of separation drops to 8. If the fraction is slightly increased to 0.03% it falls to 5. Clustering, a social reality, persists but small worldness, a counter-intuitive phenomenon appears. Courtesy of random encounters, forming relatively weak relations, but strong enough to tie a planet together. Little things do mean a lot! Now what about those fireflies and fiery neurons?

6.2 Ready, Fire, Aim

Imagine a 200 yard stretch of forest bordering a river in Papua New Guinea. The trees are forty feet tall. The scene whilst verdant and panoramic is nothing extra-ordinary. A firefly decks each leaf but you see none. Night falls. Speckles of light dapple the stretch and interest awakens. Soon there are clusters of blinking lights as near neighbors get accustomed to their fellow flashers. The scene crescendos in a series of single solid flashes, about twice per second, along the entire stretch. Millions of fireflies have synchronized themselves into an orchestra of light in the darkness. It is a vista to rival anything that Walt Disney pulls off at EPCOT. The scientific term is terrestrial bioluminescence [19]. But let us not allow technical nomenclature to obscure an inexplicable phenomenon.

How is this possible? How is synchronicity achieved? Is there a conductor for this orchestra? The ant, we are told, has no commander. Are fireflies somehow a brighter species, that a leader emerges from their uniform ranks? And if so, is it possible for millions of fireflies to notice a single leader and be smart enough to subordinate themselves to this single command? Perhaps not. By the same token it appears unlikely that fireflies have the bandwidth to tune into all their neighbors. If there were say 10,000 fireflies the total number of communication paths between them would be 50 million. Can a single firefly monitor 10,000 chat lines in the context of 50 million traffic lanes? Unlikely. The motivation is high. The flashing is the means by which males attract females as a prior to mating. It makes sense to produce a series of single blinding flashes. That will get the females' attention. Going it alone is risky. But can a forest of fireflies produce the collective consciousness to synchronize in order to maximize mating potential? Who has that idea? Some of them or all? And how do they share that notion?

To make progress with these questions we need again to turn to the seminal work of Duncan Watts and Steve Strogatz. It seems possible that near neighbors will somehow get their flash act together. By comparison two grandfather clocks near to one another in a room have exhibited a synchronism of their pendulums, the explanation being the interaction of rhythms each transmits to the other via the floor. Let us argue then that each firefly responds mostly to the flashing of a few of its nearest neighbors. The computational burden on a firefly is now more realistic, tuning in to say 5 neighbors. The traffic lanes are now reduced to 0.1% of what they were when each firefly could communicate with any other of a population of 10,000.

A rare few fireflies might also feel the influence of a fly or two at a longer distance. A few fireflies might have a particularly brilliant flash, and so be visible to others far away, or few genetic oddballs might respond more to a fainter flashes than to bright ones. In either case, some fireflies make it possible for long-distant links to exist between evident clusters. This argument begins to present the opportunity of a small world pattern.

Watts and Strogatz carried out computer simulation using this small world architecture and repeatedly found that the insects were able to synchronize almost as readily as if each one had the power to speak to any other fly. In essence the pattern is a breakthrough in computational efficiency.

No one really knows how fireflies are 'connected'. Only a few species in Malaysia, New Guinea, Borneo and Thailand have the power to synchronize evidently. So the terrestrial light orchestra is still shrouded in mystery. But a fingerprint of computational speed and power may have emerged. Armed with that it does not seem unreasonable to ask whether our human brains, computation engines par excellence, might possess a small world architecture.

Phrenology is thoroughly discredited as a body of knowledge. But it was not entirely without purpose. The notion that the brain is somehow arranged into organs, functional building blocks each devoted to specific tasks such as memory, sight, sound, emotion and so on, is one that has usefully carried over from the hocus pocus of feeling the lumps and bumps on our skull as a pattern match with personality traits, into modern day neurological science.

The cerebral cortex is held to be the locus of our higher capabilities. This thin gray intricately folded and delicately packed outer layer of the brain, just a few millimeters thick, contains the precious neurons from among the 100 billion that make up the brain's tissue. The cortex is the part of the brain that lets us speak, perform mathematics, learn music, and invent excuses for being delinquent. It is what makes us distinctively human. And it is indeed organized into something like a set of organs. MRI scans are ways of detecting neural activity based on oxygen content in the blood flow patterns around the brain. They can therefore act as a lens into the modular decomposition of the brain relative to various tasks in which we are engaged, e.g. responding to a verbal command, recognizing a taste, or recalling a friend's address.

Squeezing 100 billion neurons into a 3 pound lump inside the skull seems far fetched but not when you consider that you can get more neurons into a thimble than there are people in the United States. Crudely, a neuron is a single cell with a central body from which issue numerous fibers. The shortest of these called dendrites are the cell's receiving channels. Longer fibers known as axons are the transmission lines for the neuron. Most of the neurons link up with near neighbors within the same functional region. Signals from axons are received by dendrites and so neural activity is myriad signaling along these channels of which there are hundreds of trillions. The brain is interwoven like nothing else we know. It is complex. But is it simpler than it seems?

Whilst functional regions are in effect clusters of neural connections. The brain also has a smaller number of truly long-distant axons that link brain regions that lie far apart, sometimes even on opposite sides of the brain. Consequently the human brain has many local links and a few long-distance links, something that starts to resemble a small-world pattern. Research has shown that the degrees of separation in a cat's brain is between two and three, identical with a macaque brain, while at the same time regions are highly clustered. So it seems that what is true of social networks is also true of what Mark Buchanan charmingly calls a thoughtful architecture [20].

Some biological advantages of this small world architecture in the human brain are compellingly clear. If you accidentally hit your thumb with a 5 pound hammer instead of the intended nail, several things happen in coordinated fashion. You drop the hammer, draw the offended thumb to your mouth and suck it, let out a scream, and do a jig. At least that's what we do! Several parts of the brain are called upon to engage this kind of bodily function and that orchestration is achievable only because of the long-distance links that tie the various clusters responsible for separate actions together. A second advantage lies in the fact that brain clusters provide huge redundancy so with the wear out and/or fall out of neurons the functional blocks can still perform their functions. Even if functions are degraded or rendered impotent their separation means each block can still function so that loss of speech understanding does not necessarily mean loss of memory or the ability to make future plans. Even if communication links are broken between say the visual cortex and the hippocampus, that could result in a slight degradation in short-term memory of visual information, the small-world architecture takes care of that by providing alternate longer less direct routes. It's as if people remain neighbors even though gulfs well up; they simply use go-betweens. After all, these neurons are all in it together, all 100 billion of them!

There's one final thing to say about the brain's magic imparted by this thoughtful architecture. After construction and coordination comes consciousness. No-one knows where this resides. It is one thing for neurons to be ready and to fire, they deserve their 5 millisecond reset time having unloaded, it is another thing to take aim, to say, "I am a conscious human being, I am me, and I am unique". How does this come about? The orchestration of billions of neurons might be addressable in terms of small-world architecture and synchronicity, but how does this self-organization at the *many* level lead to self awareness at the *one* level? State of the art research seems to suggest that there are connections between the two and it is all a matter of the simultaneity of the many and the one, the ultimate paradox of complexity.

Thinking of consciousness requires us to address two aspects: conscious states and conscious organization. Consider the scenario of being a student in a class room and briefly, while being unengaged by your instructor, you glance though a window to see someone running toward your building. What do you make of this? More particularly what does the brain make of this? It engages in the generation of multiple states: of pattern recognition, movement detection, context setting, generation of emotions, awareness of sounds, selection of possibilities. All at the same time and all concluding in a single integrated picture.

This is made possible by neural synchrony: the coherent engagement of neurons and in many regions of the brain and *at multiple levels* into one overall pattern. Research has found that when the brain is confronted by two distinct views, of some simple patterns, neural activity is not synchronized when the patterns are seen separately but when they are made to merge and become one pattern, neural activity is synchronized. It is synchrony that creates conscious integration. Moreover in synchronous movement individual neurons maintain a subtle but defined lead or lag behind the group's average firing so that the whole orchestration is information rich in what it provides to upper echelon neural circuitry. It's the equivalent of hearing the orchestra and the violins and the flute. The one and the many.

One of our paradoxes in coming to an understanding of the term complex is that complexity is far simpler than it looks. Behind the apparent confusion lies a hidden order. So this proves in the small world architectures of Watts and Strogatz. That order can be summarized by tightly-knitted clusters and random connections between these. It's as if the randomness gave rise to the order we find in the computational efficiency and resilience that instances of these architectures produce, e.g. finding female fireflies and mind magic. This is encouraging. We turned the paradox to our advantage. Let's not leave it there. What else? Is it possible that order lurks behind pure chaos? Are there invisible forces at work to shape our lives, our technologies and our environment regardless of happenstance, uncertainty and accident? Can there exist a design presence that steers a course while chance itself holds sway? This is not merely an academic question or idle philosophy on our parts, though we do take great enjoyment in posing these questions for their own sake.

There are many instances of networks that exist in the real world that simply do not conform to or are shaped by the forces that give rise to the small world architectures we have thus far enunciated. Inspired as we were to abandon our stereotypical networks of electrical grids and social circles to go on and discover new varieties of this key notion, the firefly flocks and neural networks being prime examples, we can find many more examples – river, networks, air transport networks, the Internet, and the chemistry from which we humans are made. None of these exhibit the pre-existing clustering that our small world examples have required. These are all products of two forces – growth and chance. Who knows when it is going to rain, how heavily and where? Water that fills our rivers. Who decides what airports there shall be served by what kinds of aircraft traveling to who knows where? Planes that fill our skies. What determines which servers will attach to the Internet, publishing material to the Web and providing access to countless millions? Servers that thrill our surfers. As much as we might imagine we have an involvement in these things, in no way can we say we determine outcomes. These are governed by growth and chance. Do these dance? And is there a discernible choreography?

6.3 Snowballs and Seesaws

What do the great Mississippi, the Internet and a computer game, inelegantly named diffusion limited aggregation (DLA), have in common? At face value the answer is surely nothing. But that's because we are looking for a simple explanation. To find what we are looking for we must subject ourselves to paradox, and find the complexity in each of these systems.

The Mississippi River stretches 2315 miles from its source at Lake Itasca in the Minnesota North Woods, through the mid-continental United States, the Gulf of Mexico Coastal Plain, and its subtropical Louisiana Delta. Its river basin, or watershed, extending from the Allegheny Mountains to the Rocky Mountains, including all or parts of 31 states and 2 Canadian provinces, measures 1.81 million square miles, and covers about 40% of the United States and about one-eighth of North America. Of the world's rivers, the Mississippi ranks third in length, second in watershed area, and fifth in average discharge. That's the big picture. How it got formed is largely by accident.

Over however many years, the clouds gathered, the rains came, the ground washed away and the mighty Mississippi began to take its shape. We do know gravity played a part, that's why the rain comes down! As the soil erodes by the washing of rainwater, channels are formed. This has a positive feedback effect in enhancing that flow of water. Grid lines are carved on the earth as fingerprints of the rain's reign, the watermark of myriad deluges. An invisible force is at work influenced by gravity and history, that of what has gone well before will be welcomed back again. What is the imprint of this force? To answer that question we need an imaginative leap into data capture. What is the relationship between sectional lengths of the river and the amounts of water these drain? Why should we bother to ask this question? The reason for that is because the imprint of the Mississippi bears marked similarities with many other great river basins. There exists a pattern in the formation of river systems and this cannot be explained by comparing the details of their environs or history. But it can be accessed by this imaginative question.

This data conforms to what mathematicians call a power law, or what engineers call log-log. And it is this power law relationship that holds true for all river systems. It becomes attractive to think of this as the architecture for river systems, a design whose architect eludes us, without faith.

We cannot improve upon Mark Buchanan's eloquence: "The real importance of the power law is that it reveals how, even in a historical process influenced only by random chance, law-like patterns can still emerge. In terms of this self-similar nature all river networks are alike. History and chance are fully compatible with the existence of law-like order and pattern."[21]

Does this power law have ubiquity? Can it be that this architecture shows up in the Internet and this clumsily named computer game?

The Internet has its origins in many ideas including a DoD need to build resilient infrastructures, as well as the need of many researchers to share information reliably and efficiently, initially confined to California. Its origins are however of far less importance than its history, as remarkably brief as this is. That history has been governed by spectacular growth and inordinate randomness. No one determines the Internet even though its protocols are well published standards that are fully-obeyed. Is a pattern possible for such a system? Consider the relationship between the number of nodes in the Internet and the number of links these nodes possess to other nodes (indicated by the number of routers located at these nodes). Once again this relationship conforms to a power law – it is log-log.

Apparently rain falls where it will causing rivers to flow where they will, and routers sprout wherever they will creating information flows where they will. But in both cases their will be done according to a higher power law.

What about the DLA game? Imagine a blank screen and an insignificant anonymous object drifts across. A second one does likewise, both appearing and traveling perfectly at random. If they bump into one another they stick together. If they miss, they carry on their random walks perhaps disappearing from view. Millions of these objects appear over time. A figure appears. It should, according to our intuition, be an anonymous insignificant blob – an aggregation of myriad identical objects. But there is a pattern,. It is tentacled, from which we infer that it's hard for new objects to get to the centre of this non-blob. The tentacles are self-reinforcing. More than this they are self similar. What mathematicians call fractal. Just like the river systems. Our power law just won't go away! Growth and chance keep it alive and well. It shows up in the most surprising places. In our body chemistry, one or two specific molecules take part in several hundred chemical reactions involved in the bacterium's metabolism, whereas many thousands of other molecules take part in only one or two reactions. The distribution of molecular interactivity against the number of molecules with a given interactivity is yet another power law.

Given its irrepressible nature, might we dare to ask 'Does the power law provide us with a network signature? What is this like? And what rules are in operation that create or govern these 'power networks'?

Though a little naïve, it is hard not to equate the power law with 'To the victors the spoils' That is how power operates right? A colloquial expression for this is 'the rich get richer'. As much as one might despise this it seems to be a law of the universe. Is it the power law? Barabasi and his colleagues [22] answer this for us by their experiments with 'preferential attachment'.

Consider a green field situation – several nodes and no links. Gradually links are added entirely at random. With some new links come new nodes as they enter the growing network. Inevitably a few nodes will gather a few more links than others. Now consider that as new links are added they have a preference towards connecting to nodes that already show a preponderance of links to other nodes. The process of adding links is still random but the probability that these now attach to the more popular nodes is slightly increased. With such a set of rules these experiments produced a similar pattern repeatedly. What is more, the architecture of these patterns always conformed to a power law with a few nodes acting as powerful hubs and myriad nodes having relatively few links. The numbers of nodes plotting against the population of links that these nodes support falls off in log-log fashion.

We observe this phenomenon in many walks of life and instances of science. A power law fits the number of non-executive company directors with a precious few holding more than 100 offices and very many just one or two directorships. It is clear why. Corporations need savvy to inform their strategic planning. So much of this can come from non-executive directors. The ones that are most coveted are the ones that are already popular with companies, i.e. who are already serving in many capacities. It makes sense. Each corporation gains via the wildly popular non-executive director the enhanced experience that person is gaining courtesy of serving on several boards already. These folk are in effect conduits of corporate knowledge around the landscape. Conduits embodied as hubs. It is the old boys' club writ large.

We see the power law in sexual-contact networks. Inevitably a few people are more sexually active than others in terms of the partners they have. There are forces at work here. With success at gaining new partners comes an acquired skill to gain yet more. With more partners gained comes the need to practice that skill more extensively in order to keep up a good image. With that motivation comes more skill and more partners. It is a cycle that Pete Senge would call a reinforcing loop and characterize by a snowball rolling down the side of a mountain potentially producing an avalanche. A hit song from Queen [23] could not be more apposite to capture this momentum –

> *I'm a rocket ship on my way to Mars*
> *On a collision course*
> *I am a satellite*
> *I'm out of control*
> *I'm a sex machine ready to reload*
> *Like an atom bomb about to oh oh oh oh oh explode!*

This type of network also has the small world property. It is another flavor of small which causes us to differentiate between the Watts & Strogatz variety and that of Barabasi [24]. The former has been termed egalitarian and the latter aristocratic. Examples of the nouveau riche are the in-demand non–executive directors and the sexually prolific. A more obvious example are the wealthy themselves. Here is quite literally a case of the rich getting richer. Money flows between people are essentially transaction based – you give me work, I give you money. You give me money, I give you goods. This works for all of us alike. But money in return for time or goods is very limiting. It is an activity that characterizes egalitarianism. The real power comes when money works for you [25]. This involves risks but it carries rewards. Big risks carry huge rewards. Two things characterize rich thinkers. First, they put less value on their money because they have so much of it. Risks are reduced accordingly. But risks are further reduced by focusing the time that is not spent working for money on being smarter about what will work and what won't. Investment to the rich does not equate with gambling by the poor. There will always be risks since uncertainty rules but you can minimize these by investing your time wisely.

Is there an end in sight to these gains? Does reinforcement continue endlessly? Or are there limits to growth as Peter Senge found. Does the snowball meet any obstacles that can prevent the avalanche? Is a seesaw in sight?

Left unchecked the air transportation network in the USA would become aristocratic by nature with the major hubs being of course Atlanta, Chicago, and Dallas. The hub and spoke system much favored of airlines serves their needs to carry as many passengers as possible wherever they want to go, provided it is via their hubs. It makes economic sense for them to concentrate resources and facilities at major airports at the inconvenience to passengers of switching flights and layovers. But the 80 million passengers that pass through Atlanta's Hartsfield Jackson International Airport annually represents a limit to growth. The airport is often running beyond maximum capacity and when bad weather shows up, not only in

Atlanta but at connecting cities, life gets hairy. Passengers vote with their feet which explains the growth in regional airports, smaller aircraft and point-to-point travel. People are not electrons. There is a distinct difference between atoms and bits [26]. And whereas there appears no limit to how many web sites can point to and be pointed at by others, this is not the case for mere mortals with luggage and a persistent need for burgers, bathrooms, and beds.

So there are balancing forces that will arise to keep in check the rich gets richer snowball. The interesting thing for us to consider is how these network architectures, aristocratic and egalitarian, can switch. What are the factors that determine this and what are the consequences for people, corporations, species, and technology systems in terms of reliability, security, safety and resilience?

6.4 Significant Others

The persistence of the small world architecture is impressive. That it comes in these two flavors is also charming. Both types emphasize the paradox of revealing a hidden order to apparent chaos and of providing a simple elegance to what otherwise seems immensely complex. Who can be satisfied though with leaving matters there when the urge to find deeper meaning through higher order patterns has been stimulated by successes thus far? Isn't it the case that we are in a process of understanding significance, of meaning itself?

In the egalitarian networks the significance lies in the weak links, another extraordinary paradox. In the aristocratic network, more evidently the significance lies in the (super) hubs, also known as the vital few (compared to the trivial many). What exactly is the nature of this significance? That of course depends on the real world situations whose apparent disorder and complexity is elegantly captured by small world architectures.

In the case of the world's ecosystem, aristocratic small world structure is a natural source of security and stability. Yet the super hubs or 'keystone species' represent crucial organisms the removal of which might bring the web of life tumbling down like a pack of cards. Removing even 20% of the most highly connected species fragments the food web almost entirely splintering it into many tiny pieces doing untold permanent damage to a web of life. Culling of one species send out 'fingers of influence' that in a few steps touch every last species in the global ecosystem. Strong links between species set up the possibility of dangerous fluctuations therefore since the vital few are the vulnerable feet of an aristocratic giant. By sharp contrast the weak links between species act as natural pressure valves in communities. The weak had once again gone unnoticed since our concentration was focused on the vital few hubs. Paradox demands wisdom and complexity often finds us lacking in that department.

A smart David spots the temple of his opponent. An agile youth unencumbered by heavy protective armor for which no need is foreseen casts the first stone. And it is enough.[27] The lesson for us is to find the simplicity behind the complexity, recognize what is significant and, in the case of food chains, be smarter about what we can and can't cull.

Resilience, the ability to withstand major disturbances and quickly restore order is now a matter of architecture in the face of specific threats. Random networks, despite their redundancy, fall apart quite quickly in the face of an uncoordinated attack. The aristocratic network, like the real world Internet, falls apart gracefully under random attack, and doesn't suffer catastrophic disintegration. But the very feature that makes an aristocratic network safe from random failure could be its Achilles heel, in the face of an intelligent assault. As far as the Internet's wholeness or integrity is concerned, the destruction of 18% of the most highly connected hub computers serves to splinter the network almost entirely into a collection of tiny fragments.

Complexity science is the latest of King Science's courtiers but its arrival might spell the end of that monarchy. Reason has been the prime mover behind all science, the force that swept away superstition and dogmatic religion. Yet scientific method's exploration of this vast phenomena of complexity has revealed a new simplicity, a collection of elementary paradoxes that defy reason! Can science keep its principles in the face of this usurper whose initial disguise was the clothing of its own king? Or can science safely abandon its principles, in a principled way, and so extend its reign even to beyond the borders of unreasonable paradox? This is the battle now for science, one in which its only means of achieving victory is by means of surrender!

7 Paradoxical Systems

The predominant thought pattern of systems engineering is that of *choice*. Basically SE is excellent at defining a solutions space given a tenable set of constraints. What engineers are superb at is identifying candidate solutions that lie within that space, these being forthcoming on the basis of the merits they proffer relative to the criteria imposed or implied by the stakeholders. The final piece of this process, given a solutions space and a set of candidates, is the selection of the best candidate, or optimal solution. The notion of choosing, from a set of candidates, is a powerful thinking pattern, continually reinforced by the SE paradigm.

When it comes to soft systems methodology (SSM) we are stopped dead in our tracks by the multiplicity of choice, each one of which appears to take on a personality of its own demanding unique attention and requiring that no other be chosen above any one. Here is a case of damned if you do and damned if you don't. We stand at the doorway of paradox. The practitioners of SSM know full well the simultaneity of tenable viewpoints and accept the role of facilitating debate amongst the stakeholders to arrive at a course of executive action that is both systemically desirable and culturally feasible. That arrival may very well call for the abandonment of all the original viewpoints, accepting instead the emergence of fresh thinking as part and parcel of the SSM approach. In the mix, contradictions and controversies are dealt with, the force of consensus being mobilized as a product of the stakeholders' emerging sense of community, and therefore an acceptance of collective wisdom (not groupthink) that appreciates the initial set of STVs, has a maturity that knows which to abandon why and when, and a fundamental change in thinking to envision new possibilities and new strategies for realizing these. It is as though SSM transforms the multiplicity of choice into a consensus that at the outset was never a choice!

Latterly, what complexity has done is to reveal hidden threads of interconnections between hitherto remote and apparently disconnected entities and to show us how insignificant changes in any one of these can have tumultuous effects on others. In systems engineering the choices were straightforward and inert to being chosen. In

SSM the choices contended with each other each forcing themselves on the decider, to the exclusion of others. In complexity the choices are far from obvious and when they do come into the light they are massively interconnected, making it a case of choose all or none. What we have witnessed as science has progressed is a gradual erosion of the very reductionism which has powered the science revolution. It is indeed unusual for that which has wrought a revolution to secure it. However it is inexplicable that this property to be so belated, in the order of half a millennium! And now comes the final nail in the coffin!

Our scientific knowledge has brought us finally to the realization that in nature and in design our ultimate challenge to the intellectual pursuit of all mystery lies in an apparent abandonment of reason. Here we confront paradox itself. In its most generic form, a paradox is a stark choice between two clear alternatives – systems engineers would love that - but the existence of both choices denies the plausibility of either. This is not mere contention. It is the rational denial of choice based on the very existence of the same.

In the rest of this paper we seek to better understand, by exploration of an eclectic collection of exemplars, why paradox exists, what its existence means, and how revolutionary thinking, of a kind that parallels anything we have seen throughout mankind's history, must be perpetual in our world of systems.

7.1 Energy

In his brilliant book *Hot, flat and crowded*, Tom Friedman describes the Dirty Fuels System. This system has powered the industrial revolution and is indispensable to the western world for maintaining the extensive infrastructure and accustomed lifestyle of hundreds of millions of people. Naturally this system is coveted by the rest of the world and is accordingly being widely adopted and deployed in developing nations, principally India and China, that have seen the virtues this system has bestowed in the past. It is a system that has fueled growth, delivered prosperity, and affords huge influence to a nation state in the world's eyes.

However, the first name of this system heralds a note of caution. The system is seen as hellish since the fuel comes from below (the earth's surface) - in the form of oil and coal, pollutes the earth, and is exhaustible – no future in hell (other than an unwanted one). By contrast the heavenly fuel comes from above, in the guise of wind and solar, does not pollute (has a purity to it), and is inexhaustible (at least until the sun is extinguished). This dramatic comparison serves at least to draw attention to the now incontestable fact that the dirty fuels systems is bad – for all of us. It is especially bad if it is going to be adopted by all of us. However, how can anything be so bad if it has brought so much good?

That question is bound to be asked by those now wanting to adopt that system so that good might come to them, and is inevitably being asked of those for whom the system has brought much good but who are, ironically, the ones now saying that the system is bad. How would you feel if you were one of those to whom good had yet to come, but was now clearly within reach via the 'Dirty Fuels System', and was being told by those to whom much good had come that this is a bad system? It would tax your trust would it not, and possibly make you more resolute in choosing 'hell'?

This particular Gordian knot [28] is not easily sliced, and it forms the basis of a superb exemplar of a paradoxical system. Imagine a unit of energy consumption to be an Americum. This represents a group of 350 million consumers with a per capita income in excess of $15,000 and a growing penchant for consumerism. The USA is 1 Americum. For many years there were only two Americums in the world: the USA was one, the other in Europe. There were small pockets of Americum-style living in Asia, Latin America and the Middle East. Today Americums take shape all over the planet: China has given birth to one and is pregnant with another; India has one now and one on the way, due by 2030; another is forming in the Pacific rim, and yet another in parts of South America and one more in the Middle East. By 2030 we will have gone from 2 Americums to 9. These are America's carbon copies. The problem is that the filing system doesn't have room for all. Friedman quotes Jeff Wacker [29] as follows: "Our prosperity is threatened by the very foundation of that prosperity (the nature of American capitalism). We have to fix the foundation before we can live in the house again. China's foundation cannot be the same as the same foundation we built America on. And America can no longer be the same".

The problem is that we do not know what the new foundation is; and there's nowhere else to live while the old foundation is being changed. My question is, "How can SE, SSM and/or complexity science help us cut the knot?"

7.2 Geopolitics

Commentators refer to John Adams and Thomas Jefferson as 'the voice and the pen'. For sure John Adams knew to enlist the services of Thomas Jefferson in drafting the Declaration of Independence while he himself always rose to his feet at the Continental Congress with unbridled passion and convincing authority to ensure unanimity among the colonies in adopting a revolutionary posture. By the same token, Jefferson knew full well his own limitations as a public speaker preferring the solitude of his library and private moments of quiet inspiration to lend his unique skills to the revolutionary effort. And who can deny the magic of his prose as the ink flowed onto the pages with irresistible fervor and undeniable unction.

"When in the Course of human events it becomes necessary for one people to dissolve the political bands which have connected them with another and to assume among the powers of the earth, the separate and equal station to which the Laws of Nature and of Nature's God entitle them, a decent respect to the opinions of mankind requires that they should declare the causes which impel them to the separation.

We hold these truths to be self-evident, that all men are created equal, that they are endowed by their Creator with certain unalienable Rights, that among these are Life, Liberty and the pursuit of Happiness. — That to secure these rights, Governments are instituted among Men, deriving their just powers from the consent of the governed, — That whenever any Form of Government becomes destructive of these ends, it is the Right of the People to alter or to abolish it, and to institute new Government, laying its foundation on such principles and organizing its powers in such form, as to them shall seem most likely to effect their Safety and Happiness. Prudence, indeed, will dictate that Governments long established should not be changed for light and transient causes; and accordingly all experience hath shewn that mankind are more disposed to suffer, while evils are sufferable than to right themselves by abolishing the forms to which they are accustomed. But when a long train of abuses and usurpations ..."

It is borderline sacrilegious to then say 'and so on'. These are not mere words but invocations to the soul of every man ... of *all men*. At the time it was well accepted that *all* did not mean *all*. Almost a century later, in continuing pursuit of that more perfect union, *all* began to mean rather more than it once had ... and yet more remained to be done. Jefferson's words have marched on with time embracing more minds and enfranchising more of us. They have torn down walls and infected many with thoughts of a better life. The American dream is not confined to the Continent in just the same manner that Jefferson himself foresaw America would not be confined to the eastern seaboard on which its fledgling from first saw the light of day. The words though few and simple are limitless in their scope. But are they? If growth is the vehicle for the advancement of these words, what of the limits to growth? If words are power unto themselves, what are the limits of power?[30]

The expected destination of all men being created equal is a middle class earth that cannot be sustained on planet Earth. Here is the inevitable geo-political paradox: if these words are true then their truth cannot be denied to those for whom they have not been true – they need to be made true; the application of these words lifts people out of despair and sets them on a road to prosperity; the sharing of that prosperity – by the means that has thus far provided it to all for whom the words are true must lead to despair for all.

Is this paradoxical situation beyond the scope of SE? Beyond the scope of SSM also? And, even, beyond the scope of complexity science? If this be the case, is it not just too much to hear of their impotence in this matter given that these very servants of science, by their application to technological, sociological, ecological, and political endeavors have brought us to this quandary? Are they, having got us in this mess, now to wash their hands of it? (Even though they continue to practice unaltered?) Is it conceivable that an exhibition of humility among these problem-solving giants, rendered by the introduction of paradoxical thinking - the nemesis of science, might be the wicket gate that leads from this world to that which is to come?[31]

7.3 Information

The history of computing begins inevitably with man, the first computer whose productivity soared with the invention of the place value systems and was enhanced by devices such as the abacus. Automatic engines came much later onto the scene but as in so many other developments a power law applies to the growth in capacity, productivity, and innovation of both man and machine. Calculation has given way to information for its own sake and latterly the strive is away from data towards knowledge and wisdom. Some suggest that man plays god by making machine in his own image via artificial intelligence. Today computing has even taken on an ethereal nature disappearing as it were from tangible machines into mysterious clouds. Users are uncaring of the location of computers, software and data as long as their needs are met in timely and reliable fashion on whatever fashionable boxes they attach to their person. The opposite is true for cloud providers who must locate vast server farms near sources of power and cooling and who are continually frustrated by security and privacy relative to the geography that applies to specific clouds.

One thing is for sure: that the information technology and energy technology media (or infrastructures) must come together (some refer to this as cyber-physical ecosystems). Only in that way can the energy crises, the geo-political challenges and the information futures be assured. That the way to this convergence is strewn with paradoxical systems must be regarded as a necessary blockage to the promised land and a sure sign that this is the only way, one guided by breakthrough thinking.

8 Principles and Their Principled Violation

Science is founded on reason and paradox appears to call for the abandonment of reason. Yet paradox, we assert, is the necessary road block to conventional wisdom and a portal to breakthrough thinking. As with any other road block you can ignore it, you can go around, under or over it, or you can find a different way to a different destination. This paper says when you come to the paradox you must treat it as a sign post that your current thinking, that brought you this far, will not work in future. If you're prepared to do that then the way will open up to where you want to go, and, respecting the paradox as a tutor, will lead to fresh understanding. In the end you don't abandon reason you jettison failed understanding and you add to the stock of wisdom.

A second rock on which science is built is that of principles. These are not easily formed and once established are seen as immovable and immutable. They serve as our fallback in all tricky situations. Were these to vanish, so the argument goes, all would be lost. What happens then when we are forced to abandon our principles? Are we acting in an unprincipled manner? Perhaps, but not necessarily. Groucho Marx, that most irreverent of comics, once said, "I have principles, of course I have principles. And if you don't like them, well, ah well, …. I have other principles". Hilarious. But many a true word spoken in jest. Let's explore this.

8.1 Declarations

In his book American Creation (Ellis, 2007), Joseph J. Ellis reveals the centrality of paradoxical thinking in the founding of the Unites States of America. It is prominent in the Declaration of Independence, the Constitution, and the necessary actions of the founding fathers.

Like their contemporary Rousseau, the members of America's revolutionary generation rejected the long-held conventional wisdom that political sovereignty must be singular, indivisible, and vested in a single location. The Constitution created multiple, overlapping sources of authority in which the blurring of jurisdiction between federal and state power became an asset rather than a liability, transforming the idea of classical "sovereignty" and reserving it in a holistic repository—the people.

'Problematic', 'blurring', 'elusive', 'endless', 'multiple' and 'overlapping' were defining characteristics of this new generation, not in any paralyzing or disingenuous sense but, by the sharpest of contrasts, in a vitally paradoxical sense, that captures the imagination and brings hope for an expansively endearing and enduring empire of liberty. It is, after all, notably uncommon for the men who make a revolution to secure it as they did.

Many Americans fail to recognize the profound paradox expressed in two of our most cherished documents. The Declaration of Independence locates sovereignty in the individual; in some ways it depicts government as an alien force, and makes revolution against oppressive government a natural act. But, the Constitution locates sovereignty in a collective called "the people," values social balance over personal liberty, and makes government an essential protector of liberty rather than its enemy.

8.2 Washington

In the winter of 1777-78, George Washington selected Valley Forge as his encampment because of its abundant food supply, which he wanted to deny the British troops ensconced in nearby Philadelphia and claim for his own troops. Unfortunately, the bulk of the local farmers preferred to sell their crops to the British army in Philadelphia. A veritable caravan of grain and livestock flowed from the very productive countryside into the city to feed British troops. Women and children drove the wagons to minimize the likelihood of arrest by American patrols.

At least some of Washington's peers viewed his plight sympathetically. "I am the more chagrined at the want of provisions to which I am informed your Army is reduced as I believe it is partly owing to the boundless Avarice of some of our Farmers, who would rather see us ingulphed (sic) in eternal Bondage, than sell their produce at a reasonable price". So wrote William Livingstone, the governor of New Jersey.

Washington was caught between two equivalently incompatible courses: confiscate the food for his starving soldiers, alienating the very people the American Revolution was intending to defend, or maintain the revolutionary principles while watching his army dissolve. What a dilemma! What a paradox! What to do?

Washington took decisive action in multiple roles. As commander, leader, judge and problem solver, he chose to inflict short-term suffering to achieve a long-term outcome—free American people living according to American principles. He put his soldiers first by ordering the Army to intercept and confiscate the wagons, horses and food, but to leave the women and children driving the wagons unharmed. The Army arrested and hanged the ringleaders of the illicit traffic in public executions designed to make a statement, and shot men carrying food into Philadelphia, leaving their bodies on the road as warnings.

Famous for both his personal reserve and scrupulous ethics, Washington had to fill another, perhaps far more difficult role—arbiter of his own conscience. Even here, circumstances prevented doing all he surely would have wished. The General ordered the Continentals who confiscated food to pay for it with certificates of debt. Unfortunately, these were essentially worthless to the bereft farmers and their families.

Washington simultaneously exercised and was blessed by wisdom, in that he knew the people's necessary suffering lay on a road to a destination they themselves had determined, yet had lost sight of: life, liberty and the pursuit of happiness. His difficult choices also exposed a new development in the strategic chemistry of the war. It became clear for the first time that the American Revolution, while still and always a struggle for independence, was also a civil war for the hearts and minds of the American people—the two contests were closely connected. Washington had previously viewed the war as a conventional contest between two armies, so he had focused on sustaining a fighting force in the field that could match and eventually defeat the British army. Conventional wisdom says "the longest purse" wins the war; paradoxical wisdom sheds new light. The new variable was not money but allegiance. Even the longest purse could not defeat the deepest resistance.

When faced with the challenge of how to deal with the farmers, Washington recognized that the real problem lay not with the farmers, but with how he viewed—and must conduct—the war. This recognition led the General to a profound change of strategy—from simply fighting the British army to "covering the country" by deploying the Continentals and local militia units as roving police who controlled the countryside. This essentially defensive or delaying strategy made control of the countryside as crucial as winning battles. Had he not faced the reality of paradox, he would have missed the opportunity to both shift the paradigm and create America.

8.3 Jefferson

On a single day in 1803, the United States doubled in size, adding what is now the American Midwest to the national domain—all the land from the Mississippi to the Rocky Mountains and the Canadian border to the Gulf of Mexico. This tract of land was the most fertile of its size on the planet. It made America self-sufficient in food throughout the 19th century and the agrarian superpower of the 20th. The fact that the man who made this monumental executive decision was on record as believing that the energetic projection of executive power was a monarchical act only enhances the irony; and irony is a close cousin of paradox, with humanity often playing the role of unwitting victim.

Jefferson saw the Louisiana Purchase as vital to the nation's future. His response to James Monroe's letter in response to a slave revolt said, " ... it is impossible not to look forward to distant times, when our rapid multiplication will ... cover the whole northern, if not the southern continent with a people speaking the same language, governed in similar forms, and by similar laws ...". Jefferson's vision rested on two assumptions: that the land would be settled (by multiplication and migration) and not conquered; and, that in the interim, the most docile of sitting tenants should be the Spanish, whose imperialistic enterprise was a thing of the past (after the defeat of the Spanish Armada by the British in 1588).

Jefferson was also emboldened by the military strategy lessons Washington had learned when it came to repelling a mighty invading force bolstered by a powerful economy, but setting foot on soil thousands of miles from home. For this republican to honor the breakthrough thinking of a federalist he had come to despise and almost think of as a traitor gives credence to his statement: "We are all republicans; we are all Federalists." He pulled off the Purchase in the manner of a Caesar, negotiating directly with Napoleon, who was acting against the advice of his ministers. Concerned the emperor could change his mind, Jefferson had to act quickly, but lacked the authority to act alone. Or did he?

The Constitution's inherently ambiguous definition of executive power can expand and contract like an accordion, making the music required by different historical contexts. Forced to choose between losing Louisiana and abandoning his Constitutional principles, Jefferson subverted the Congressional process, saying, "To lose our country by scrupulous adherence to written laws would be to lose the law itself." Jefferson had a vision of what 'our country' meant, what written laws – the letter of the law – meant, and what might be meant by something more 'spiritual,' even more fundamental than a principle. Of course in such matters, without the benefit of codification, we necessarily admit ambiguity, but for Jefferson this was less a problem than an opportunity.

This consummate republican paradoxically exercised supreme executive power to fulfill his vision of 'our country.' He thought that his action would not become a precedent, but history has proved him incorrect. Indeed, history has conclusively demonstrated that how one interprets and leverages the Constitution is perhaps far less a matter of principle than a function of one's power position and perception of the situation.

8.4 Gerstner

In the early 1990's IBM faced unprecedented difficulty. The unthinkable was in the minds of some – that IBM would go under or be bought out by an upstart. Even in their darkest hour however the reality was that IBM had unsurpassed technical expertise, to build exciting new products, and unrivalled business knowledge, to build lasting partnerships with their customers. It might even been said, paradoxically, 'Is this what got us into our troubles?' Or, 'If we have these assets and are still in trouble what's our future?' Lou Gerstner stepped up to the plate.

He announced what should have been already well known and was being leveraged by IBM's competitors to Big Blue's demise - that a revolution had occurred: Information Technology was no longer being driven by vendors but by customers!

That paradigm shift was what IBM needed to recognize and leverage if they were to survive. Accordingly they formulated four new models. A new industry model: innovate or integrate; a new business model: service-led; a new computing model: infrastructure plus ubiquity; and, a new marketplace model: an open playing field [32].

But I ask, 'What was the moment of breakthrough and how was this intuited?' My argument is that as much as IBM were being blindsided by the shift in IT driver – from vendor to customer, they were also being road blocked by paradoxes of their own making. The first was this: the more they invested in IT the less it impressed customers, (because the world had changed and IT was being customer-driven). The second was made up of these two components. That proprietary systems were good for them, it locked-in customers, but bad for customers ("we the customers" throw of the tyranny of proprietary systems, the kings of IT epitomized in IBM - and demand open architectures, industry standards, end-to-end solutions, services etc.). At the same time, open systems architecture, in the case of the IBM PC, gave Wintel its start and had done almost nothing, relatively, for IBM, for whom open appeared to be not the way. So for IBM it was a case of 'damned if we do and damned if we don't.' Perhaps these paradoxes went unobserved – I am speculating here. Perhaps they were noticed but not respected. But maybe they were articulated, with respect, and that made IBM realize more fully the dramatic shift that had taken place among customers AND that IBM still had much to offer. All that was needed was a change in thinking.

One that leveraged IBM's technical expertise and business knowledge to give the company an appropriate setting for the new wind direction.

9 Paradoxical Thinking

No one can ever be satisfied with leaving a problem exactly where it is. Maybe the problem is well defined. If so, then it is to be solved, by current or future methods. This is the approach taken by systems engineering. Maybe the problem is not well-defined. Then it is to be transformed until it becomes either well-defined or sufficiently well described that some form of action can be taken as remedy. This is the work of soft systems methodology. Maybe the problem is exotic with many threads interweaving throughout creating patternless chaos. In that case we might wait long enough to watch for patterns to emerge, in which case we can use the pattern as a means of describing the problem and hence formulate a solution to the original chaos. Or we adopt a simple stance on what might be the pattern and see if this simplified structure, that gives us our hypothesized pattern, does indeed create the chaos we originally witnessed for ourselves. This is the approach of complexity science.

Everything we have ever known concerning the existence of problems, however defined, described or presented, is action-oriented – problems simply cannot be left to themselves. To admire a problem and leave it at that is to risk castigation by scientists in particular and society at large. There can be no exceptions. Not even for paradoxes. There must then be an action-oriented agenda for paradoxes, what we might call paradoxical thinking, even if the very nature of a paradox is to produce a kind of problem paralysis, a torpor that incapacitates action, something that insists that the problem be perpetually preserved. Accordingly I close this paper by offering such an agenda for action, my perspective on paradoxical thinking.

My first point is that I choose to locate this agenda not as a body of knowledge that can be executed, to stand with SE, SSM or complexity science. Instead I offer it as a component of systems thinking. In that sense I make my agenda for action not as a doing entity but as a thinking entity, having regard for what Albert Einstein said, "In the brain, thinking is doing". I believe that a paradox IS a system and therefore doing something about a paradox requires thinking about it as a system.

Accordingly the value of systems thinking needs to be restored so that it is respected by systems practitioners, probably by the offerrings it provides to them for taking action. Thus icons such as the Conceptagon, the systemigram, causal loop diagrams (CLDs) and the system archetypes which emerge from CLDs e.g. 'fixes that fail', accidental adversaries', and 'tragedy of the commons' are the visible face of systems thinking to the systems practitioners. Seeing these icons in use by that community should be regarded as successful implantation of precious seeds the fruit of which ought to be a more horizontally integrated community, across all systems disciplines, one that will want to encourage the growth of new seeds and thereby esteem for systems thinking.

My second point is that I do not regard paradoxes as intractable problems. I do regard them as challenges to the way I have been thinking, a pattern of thought that has led us inevitably to the paradox. It is as though the paradox was a way of revealing ill-health, in thinking, and closes down future possibilities until that ill-health is treated. In other words the paradox is not the problem; the real problem lies within us, within our ways of thinking, that requires us to get out of ourselves, as it were. In some ways then there are similarities between paradox and the kinds of problems that SSM seeks to treat. Wherever paradoxes are to be found we will surely discover also one or more problems that are systemic. Systemic problems require systemic solutions. SSM itself is a systemic solution. The problems that SSM treats are likewise systemic, and SSM develops a system of 'solutions', more correctly idealized worldviews, the collective sharing of which with problem owners, by skilful facilitation grounded in systems thinking, conceivably produces executive action grounded in a holistic respect for the original systemic problem.

My third point is that whereas all forms of problem solving us force to make a choice, i.e. define solution, from all that is possible (amongst all candidates), all forms of paradox force us to make a choice that is impossible. The essence of all paradox leads us to the point of convergence of the impossible and the imperative. Since culturally, historically and reasonably one is eventually forced to make a choice, the same is true of paradox. But the curious thing about paradox is that unless you accompany that choice with breakthrough thinking that choice is nugatory. The paradox lives on and is only once and for all treated, not by the choice you make, but by the realization that it is the chooser who must change. In paradox, we are the subject matter, and it is against us that revolutions must come.

References

1. Gorod, A., Sauser, B., Boardman, J.: System of Systems Engineering Management: A Review of Modern History and a Path Forward. IEEE Systems Journal 2(4), 1–16 (2008)
2. Friedman, T.L.: Hot, flat and crowded
3. Mitchell Waldrup, M.: Complexity
4. Checkland, P.B.: Systems Thinking, Systems Practice
5. http://www.sysml.org/specs.htm
6. http://yeli.us/Flash/Fire.html
7. Dove, R.: Response Ability: The language, structure and culture of the agile enterprise, http://www.amazon.com/Response-Ability-Language-Structure-Enterprise/dp/0471350184
8. Fine, C.H.: Clock Speed: Winning Industry Control in the Age of Temporary Advantage. HarperCollins, New York (1998)
9. Brolick, E.: Presentation to University of Detroit Mercy College of Business Administration Alumni (April 13, 2008), http://www.udmercy.edu/alumni/newsletters/cba/articles/brolick/
10. Dawkins, R.: The Selfish Gene. Oxford University Press, New York (1976)
11. Kimmel, L.: Lord Nelson and Sea Power (1995), http://www.geocities.com/Athens/3682/nelsonsea.html (adapted with permission)
12. Wicked problems
13. Counter intuitive behavior of social systems
14. Gall, J.: The Systems Bible (See also systems bible and systematics)
15. Gleick, J.: Chaos: Making a new science, http://www.amazon.com/Chaos-Making-Science-James-Gleick/dp/0140092501
16. Granovetter, M.: The strength of weak ties. Amer. J. of Soc. 78, 1360–1380 (1973)
17. Watts, D.: Six Degrees. Vintage, New York (2004)
18. Watts, D.J., Strogatz, S.H.: Collective dynamics of small world networks. Nature 393, 440–442 (1998)

19. See, for example, http://www.dartmouth.edu/~dujs/2000S/06-Biolumen.pdf
20. Buchanan, M.: Nexus, p. 64. Norton, New York (2002)
21. Buchanan, M.: Nexus, p. 103. Norton, New York (2002)
22. Barabasi, A.-L.: Linked. Penguin, New York (2003)
23. Don't Stop Me Now is a 1979 hit single by Queen, from their 1978 album Jazz. Words and music by Freddie Mercury
24. Barabasi, A.-L.: Linked: how everything is connected to everything else and what it means, http://www.amazon.com/Linked-Everything-Connected-Else-Means/dp/0452284392/ref=sr_1_1?ie=UTF8&s=books&qid=1228160824&sr=1-1
25. Kiyosaki, R.T., Sharon, L.L.: Rich Dad, Poor Dad: What the Rich Teach Their Kids About Money–That the Poor and Middle Class Do Not! Business Plus, New York (2000)
26. Negroponte, N.: Being Digital. Vintage, New York (1996)
27. The Holy Bible, 1st Book of Samuel, ch. 17
28. http://en.wikipedia.org/wiki/Gordian_Knot
29. http://wistechnology.com/articles/4332/
30. Bacevich, A.: Limits to Power
31. Bunyan, J.: Pilgrim's Progress
32. Gerstner, L.: Farewell IBM

Semantic Service Search, Service Evaluation and Ranking in Service Oriented Environment

Hai Dong, Farookh Khadeer Hussain, and Elizabeth Chang

Digital Ecosystems and Business Intelligence Institute, Curtin University of Technology,
GPO Box U1987 Perth, Western Australia 6845, Australia
{hai.dong,farookh.hussain,elizabeth.chang}@cbs.curtin.edu.au

Abstract. The theory of Service Oriented Environment (SOE) emerges with advanced connectivity of the Internet technologies, openness of business environment and prosperousness of business activities. Service, as a critical object impenetrating every corner of SOE, is a hot research topic in many research domains. Software Engineering (SE), as a subject in engineering field, its researchers pay more attention to supporting advanced technologies for promoting service activities in SOE. In this paper, we draw the position in the research field of semantic service search, service evaluation and ranking in SOE. By means of the case study and literature review research approach, we discover the research motivations and research issues in this field.

Keywords: quality of services (QoS), research issues, semantic web technologies, service evaluation, service search, service oriented environment (SOE), service ranking.

1 Introduction

Service Oriented Environment (SOE) as defined by Chang et al. [1] is a collaborative, shared and open environment which provides agents with infrastructures and technologies to carry out business services. It consists of four basic components, which are agents (service providers and users, buyer and sellers, websites or servers), business activities (product sales, service deliveries, marketing or information sharing), infrastructures (networks communications), and technologies (service publishing, discovery, binding and composition) [1]. Service, as a critical object impenetrating every corner of SOE, is a hot research topic in many research domains. Software Engineering (SE), as a subject in engineering field, its researchers pay more attention to supporting advanced technologies for promoting service activities in SOE. In this paper, we draw the position in the research field of semantic service search, service evaluation and ranking in SOE.

The rest of this paper is organized as follows: first of all we attempt to explore the research motivations in this field by a case study, then by literature review, we will discover the issues in the field of semantic service search, service evaluation and ranking in SOE; next we will formally define the research issues in this field; and finally the conclusions are drawn in the end.

M. Ulieru, P. Palensky, and R. Doursat (Eds.): IT Revolutions 2008, LNICST 11, pp.111–117, 2009.
© ICST Institute for Computer Sciences, Social-Informatics and Telecommunications Engineering 2009

2 Case Study for Studying the Motivations of Semantic Service Search and Service Evaluation, Ranking Research

In this section, by means of a case study, we will analyse the research motivations in the fields of semantic service search and service ranking.

As an example, let us assume that John lives in Perth (capital city of Western Australia) and desires a horse moving service provided by a local competitive company, in order to help him to move horses from Perth to City B. From the perspective of the internet services, there are two primary categories of service search engines that can be found by John.

The first category is generic search engines, such as Yahoo and Google. For example, John can enter *"horse moving companies in Perth"* into a generic search engine (here the example is Yahoo). From the retrieved results from the search engine (Fig. 1), it is observed that most of the retrieved results do not match John's search intention – horse moving companies in Perth, and the service information is difficult to be distinguished and identified from the results. Thus, it is asserted that the performance of the generic search engine is poor in this service search case study.

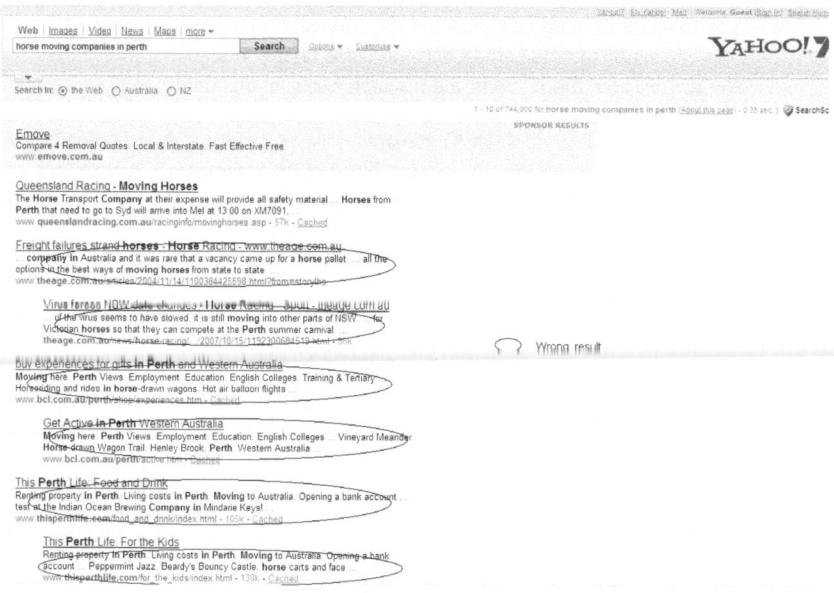

Fig. 1. Retrieved results of horse moving companies in Perth from Yahoo search engine

The reasons behind the poor performance of the search engine can be concluded as follows:

- The search engine uses traditional keywords-based search strategy without incorporating or taking into account semantic web technologies to assist the search engine to fully understand the sense of the user's query words. This causes the poor performance of the search engine in precision.

- The generic search engine is not specially designed for the purpose of service searching. As a result of this, the search process has to be carried out against a much larger information source. Due to this reason and due to the fact the search process is keyword-based the retrieved search results are not accurate and do not consider the context of the search query.
- The format of the retrieved service information is not standardized, which makes users difficult to read and comprehend the retrieved service information.

An enhanced approach is that John can access into a repository of local business directory, such as Yahoo or Google local search, online Yellowpages. These local search engines (here the example is Australian online Yellowpages) normally can provide John with two options of service search as follows:

- One option is that John can browse businesses under the *"horse transport"* category in the location *"Perth WA"*, by following the *"browse by category"* (Fig. 2). This style can provide John with more precise search results and structured service information. The defect is that John needs to follow the whole category of the website step by step, which is expensive in terms of time and effort.

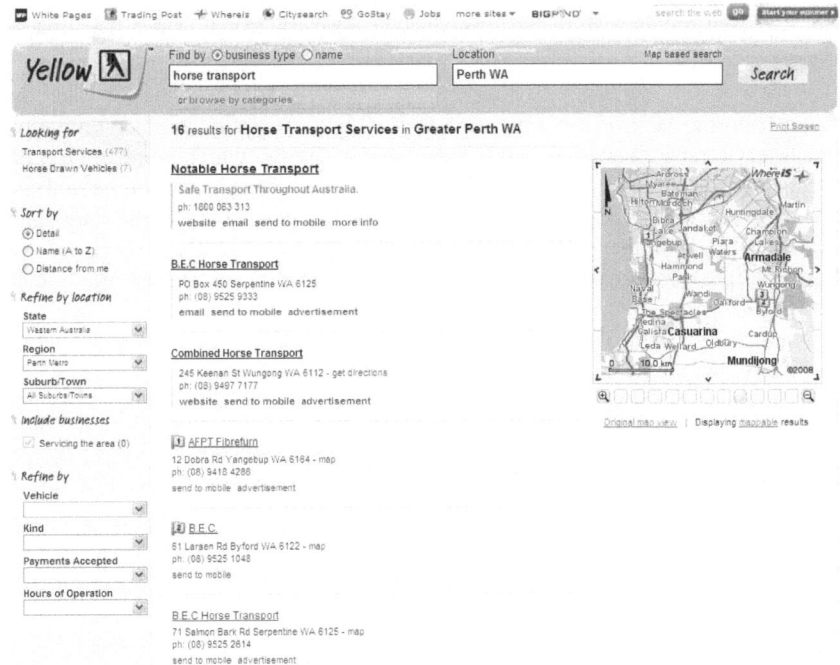

Fig. 2. Businesses under the category of *"horse transport"* in Australian Yellowpages website

- Another option is that John can directly enter "*horse moving*" into the business type box and "*Perth*" into the location box of the search engine provided by the website (Fig. 3). This can save its searching time, but this approach has its own disadvantages as well – the search engine cannot understand the user's query intention and thus returns non-relevant results. Similar to the generic search engines, the reason behind this is that the local search engine does not use semantic web technologies to help users to denote their searching concepts.

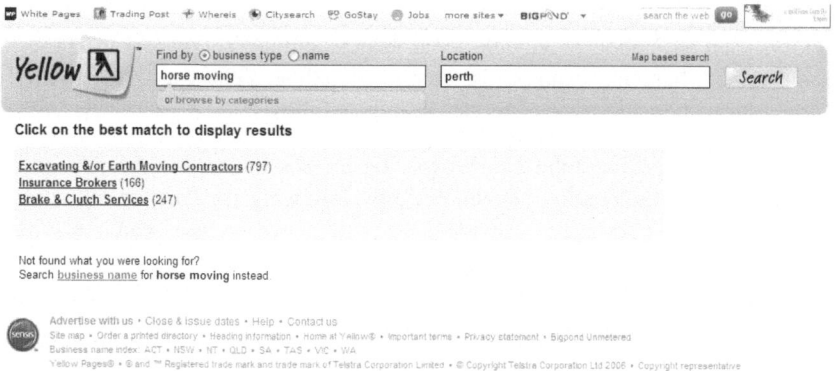

Fig. 3. Retrieved results from online Australian Yellowpages search engine based on query words "*horse moving*"

Apart from the lack of semantic web technologies' support, another limitation of the local service search engines is that John cannot find out which company has better performance from the perspective of horse moving service. The reason behind that is that these search engines do not provide user-oriented evaluation and ranking mechanisms based on the quality of services (QoS) provided by these services.

Based on the above case study, it is observed that both of the generic and local search engines are far from perfect, when searching for a given service. The research motivations in the field of semantic service search, service evaluation and ranking in the SOE environment can be concluded as follows:

- Designing multi-domains or domain-specific service search engines with semantic web technologies enhancing the search precision (and);
- Designing user-oriented evaluation and ranking methodologies for retrieved service information by means of the QoS measures.

In the next section, following the two research motivations, we will make a brief survey about the status of current research in this field.

3   Related Works

In the section, we briefly review the current literature with respect to semantic service search and service ranking and analyze their limitations.

3.1 Semantic Service Search

While there are a great number of semantic search engines being developed (e.g., SWoogle, TAP), few of them attempt to provide optimized solutions for the service search field.

Liu et al. [3] developed an e-service platform integrated with semantic search for e-service metadata. E-service metadata refers to the descriptions of e-services and providers, which is adopted to publish and to discover e-services. There are two types of metadata in the system: business level metadata – the description of e-service providers, and service level metadata – the description of basic information about e-service. The authors adopt Universal Description, Discovery and Integration (UDDI) which is a web service standard to register and search e-services. Three means for searching service and business are provided, which are find_business, find_service and XQuery. Find_business is to return a list of service providers for specific conditions; find_service is to return the information for a list of services who match customized conditions; XQuery is to query extended metadata added in a businessService list.

The limitations of the e-service search engine can be concluded as follows:

- Only one-tier (service categories-services) concept hierarchy cannot reflect the complex relationships between services in the SOE environment (and);
- There is no methodology provided for the concept hierarchy update in order to adapt for the change in service environment (and);
- The volume of its knowledge-base seems so limited that it only can be applied in limited fields (and);
- There is no ranking methodology provided for the querying results, which could lead to unorganized data structure and presentation to the user.

3.2 Service Evaluation and Ranking

While a large amount of literature focuses on evaluating quality of services (QoS), few of them study on the integration of service evaluation and service ranking system for service retrieval in the SOE.

Toma et al. [4] propose a web service ranking system based on two different ranking strategies. One strategy is to use the Web Service Modeling Ontology (WSMO) to describe the values of Non-Functional Properties (NFPs) of web services, such as QoS, Service Level Agreement (SLA). Hence web services can be ranked according to the values of user-preferred NFP. Another strategy is a multi-criteria ranking, which considers ranking multiple NFPs from three main perspectives – the user-preferred NFPs, the level of importance of the NFPs, and the ascending or descending order of services.

Gekas [2] propose a set of metrics for web service ranking. Four main categories of ranking strategies are provided by these metrics, which are degree-based rankings that calculate the percentage of fed services in each web service, hubs-authorities-based rankings that calculate the ratio between the number of incoming services and the number of outgoing services, non-functional rankings that focus on the NFPs of web service, and non-connectivity rankings that focus on the connectivity of web service networks.

The limitations of these service ranking systems can be concluded as follows:

- They are all designed for the web service environment, which cannot adapt to the broader and more complicated SOE (and);
- Despite they are all equipped with the QoS ranking methodology; none of them integrates their ranking systems with the corresponding QoS evaluation systems, which cannot ensure the correctness of the QoS data (and);
- None of them consider the factor of trustworthiness and reputation in the service ranking (and);
- None of them attempt to implement their service ranking methodologies into service search engines.

As can be clearly seen from the discussion above, research is being carried out independently in the fields of both of the semantic service search and ranking, without any attempt to integrate them together. In the next section, we define this issue formally.

4   Research Issues in Semantic Service Search, Service Evaluation and Ranking Field

In this paper, we combine our findings from the previous case study (Section 2) and our review of the existing literature (Section 3) and define the research issues in the semantic service search and service ranking field. The core research question is:

How to design a multi-domain or specific domain service search engines in SOE enhanced by semantic web technologies and QoS evaluation and ranking methodologies?

The potential research issues under this core research question are:

- For multi-domain or specific domain service search, multi-domain or specific domain service ontologies need to be designed according to the particular service domain knowledge in SOE.
- Multi-domain or specific domain service metadata format needs to be designed according to the service domain knowledge in SOE, in order to standardize service metadata.
- Conceptual frameworks of multi-agents needs to be designed to harvest service metadata from SOE according to the multi-domain or specific domain service ontologies and metadata formats.
- Multi-agent communication mechanisms need to be designed to improve harvest efficiency.
- Mechanisms needs to be designed for updating the knowledge-bases that store the multi-domain or specific domain service ontologies, in order to allow the ontologies to adapt to the dynamic change of knowledge in all service domains of SOE.
- Search algorithms need to be designed based on the multi-domain and specific domain service ontologies, in order to realize the enhancement from multiple perspectives of search engine performance.

- The conceptual framework of QoS evaluation and ranking methodologies needs to be redesigned or selected from the existing methodologies.
- Domain-specific service quality evaluation standards need to be designed according to the specific service domain knowledge in SOE.
- Mechanisms need to be designed for updating the domain-specific service quality evaluation standards, in order to allow the standards to adapt to the dynamic change of knowledge in the service domains of SOE.
- A conceptual framework for integrating the ontology-based search engines and QoS evaluation and ranking methodologies needs to be designed.

5 Conclusions

In this paper, we propose that further research needs to be carried out in the field of semantic service search and ranking in SOE. First of all, by a case study, we discuss the problems of the current generic and local search engines in the process of fulfilling users' requirements with respect to precise service search, service information standardization, service evaluation and ranking based on QoS, and so on. Based on the problems, we find the research motivations in this field. Next, by exploring the relevant literature in this field, we found that currently there is no literature on integrating the fields of semantic service search, service evaluation and service ranking based on QoS measures. Thus, we define the core research question in this field – "*how to design a multi-domain or specific domain service search engines in SOE enhanced by semantic web technologies and QoS evaluation and ranking methodologies?*". In addition, the core research question underpins ten research issues, which we have explained in Section 4.

References

1. Chang, E., Dillon, T.S., Hussain, F.: Trust and Reputation for Service Oriented Environments-Technologies for Building Business Intelligence and Consumer Confidence. John Wiley & Sons, Chichester (2005)
2. Gekas, J.: Web Service Ranking in Service Networks. In: The 3rd European Semantic Web Conference (ESWC 2006), Budva (2006)
3. Liu, D.-R., Shen, M., Liao, C.-T.: Designing A Composite E-service Platform with Recommendation Function. Computer Standards & Interfaces 25, 103–117 (2003)
4. Toma, I., Roman, D., Fensel, D., Sapkota, B., Gomez, J.M.: A Multi-criteria Service Ranking Approach based on Non-functional Properties Rules Evaluation. In: Krämer, B.J., Lin, K.-J., Narasimhan, P. (eds.) ICSOC 2007. LNCS, vol. 4749, pp. 435–441. Springer, Heidelberg (2007)

Quality Measures for Digital Business Ecosystems Formation

Muhammad Raza, Farookh Khadeer Hussain, and Elizabeth Chang

Digital Ecosystems and Business Intelligence Institute
Curtin University of Technology, Perth 6845, Australia
muhammad.raza@postgrad.curtin.edu.au,
{Farookh.Hussain,Elizabeth.Chang}@cbs.curtin.edu.au

Abstract. To execute a complex business task, business entities may need to collaborate with each other as individually they may not have the capability or willingness to perform the task on its own. Such collaboration can be seen implemented in digital business ecosystems in the form of simple coalitions using multi-agent systems or by employing Electronic Institutions. A major challenge is choosing optimal partners who will deliver the agreed commitments, and act in the coalition's interest. Business entities are scaled according to their quality level. Determining the quality of previously unknown business entities and predicting the quality of such an entity in a dynamic environment are crucial issues in Business Ecosystems. A comprehensive quality management system grounded in the concepts of Trust and Reputation can help address these issues.

Keywords: Business Ecosystems, Quality measures, Quality dynamics, Trust and Reputation.

1 Introduction

Business activities are inherently complex in nature. These activities may involve more than one entity (business entity) forming a coalition or an alliance with multiple other business entities in order to deliver objectives. Business entities, specifically SMEs, may collaborate (or form an alliance between themselves) in order to carry out these complex tasks. No single SME by itself would have the necessary resources and infrastructure in order to carry out complex activities. This may lead to the formation of a business ecosystem where more than one SME may work with other SMEs in order to achieve a pre-defined objective or goal. In such an environment, SMEs are interdependent to carry out a task [1]. Of crucial importance in such a setting is to find the appropriate partners for the business interaction along with the management of the business relationships. With widespread adoption and support of the internet and related technologies, physical Business Ecosystems can now function as Digital Business Ecosystems in order to collaborate on a virtual basis. So far, Business Ecosystems are practically implemented through several approaches, from simple concepts of coalitions using multi-agents systems [2] to a more normative form of

M. Ulieru, P. Palensky, and R. Doursat (Eds.): IT Revolutions 2008, LNICST 11, pp. 118–121, 2009.
© ICST Institute for Computer Sciences, Social-Informatics and Telecommunications Engineering 2009

Electronic Institutions [3]. The formation of Business Ecosystems, either through simple coalitions or Electronic Institutions, generally follow two approaches being either cooperative (beneficial to all the entities) or competitive (non-cooperative) [4]. Coalition formation algorithms can be classified into static and dynamic [5]. Static approaches are those that do not allow for possible changes to the membership of the coalition due to the emergence of new information while this is possible in dynamic approaches. The concept of 'dynamics' can also be found in Electronic Institutions [6]. To undertake digital business ecosystems, negotiation plays an important role in both the approaches of coalitions and Electronic Institutions. The negotiation protocols, which are crucial in the formation of Business Ecosystems, can be classified into pre-negotiation and post-negotiation [7]. In pre-negotiation protocol, the negotiation with the suppliers or providers is done before the coalition forms. In each approach of business collaboration (regardless of its nature), the underlying crucial issue is the selection of appropriate partners to ensure successful outcomes. During the negotiation phase, compromises and subsequent agreements are made on the payoffs and their distribution. However, the overall quality of the entities involved in the collaborative task execution along with the quality of the coalitions needs to be modelled precisely taking into account all relevant factors. The challenges increase, especially in a dynamic environment, where it is very important to maintain and monitor the quality of each entity along with the entire coalition.

2 Quality Related Issues

The coalition formation process may consist of two steps [8]. In the first step a task is determined and in the second step agents are selected for task completion. In a multi-agents system the agent, typically, first determines their values for all coalition (reward), second, the agent rank and select their preferences and in the end the coalition member internally distributes the expected revenue [9]. But the key concern, when forming a business alliance is the quality of the participating entities [10]. The increasing demand of users for high quality (QoS) and timely information is putting businesses under increasing pressure to update their knowledge and identify new ways for collaborating with their peers [11]. A multi-objective, optimisation, evolutionary algorithm enables an agent to choose an optimal set of agents with whom a coalition can form for a particular task, which is called a coalition calculation [12].

Here, the optimal set of agents definitely means the agents having high quality in terms of their service. It is believed to be absolutely crucial that no compromises be made in quality of the entities and coalitions themselves, which helps to address problems like uncertain future behaviour, operational breakdowns and unexpected output of the entities and coalitions. It is also strongly believe that this is only possible when entities have opportunities in selecting customised quality levels as every coalition may not need the same quality level for the business interaction in a business ecosystem. It will help to address the issues discussed in the following section.

3 Approach

The success and failure of a business ecosystem is directly linked to the quality of the entities in delivering on the predefined objectives. So far, negotiation protocols are used to enable agents to negotiate and form coalitions by providing them with simple heuristics for choosing coalitions partners. These protocol and heuristics allow the agents to form coalitions in the face of time constraints and incomplete information [4]. The crucial issue, at this stage, is not the incompleteness of information about the entities but being able to reliably determine the quality of entities for which there may be limited relevant information. In such a scenario, the quality of that entity needs to be determined through available information taking into account the semantic gap between that information and the relevant information for quality purposes. As a result, such assessments of the quality of such entities that have either little or no relevant previous information (for determining quality), can be made and this would not disadvantage them adversely, thereby solving the cold-start issue. A comprehensive mechanism, before the negotiation process, that is capable of determining the quality of such entities joining the business ecosystems would be beneficial. It is a risky decision to consider such unqualified entities for the coalition formation and particularly for the business ecosystem. This will lead to the problems mentioned in the previous section.

In today's dynamic world where the necessities and opportunities are changing rigorously, entities may find better opportunities in the future or such environmental changes may affect their quality in the future. These dynamics play an important role in the completion of tasks and the fate of the ecosystems. In order to deliver on the predefined objective, agreed upon at the time of ecosystem formation, the dynamic nature of quality at the point of time in the future needs to be predicted. This will help to prepare optimised agreements during the negotiation process [10].

Time is considered as the core factor of dynamics in quality. Other factors are dependent on the time that can be useful in predicting the quality of the entities or the entire coalition. One such time-dependent factor that can help predict the quality of entity, or the entire coalition, is the maturity of the entities. Considering the maturity of the entities or business ecosystem itself can help in prediction and also reduce the complexity in computing the future values for the quality.

4   Conclusion

The success and failure of a business ecosystem is directly linked to the quality of the entities in delivering the predefined objectives. No compromise should be made in determining the quality of the entities and coalitions themselves in a business ecosystem. Measures should be developed to deal with problems like uncertain future behaviour, operation breakdowns and unexpected outputs of the entities and coalitions. It is only possible when entities have opportunities in selecting customised quality levels.

To deal with the issues discussed in previous sections and to enable such quality control mechanisms, the role of trust and reputation cannot be ignored. Trust and Reputation are well known in creating environments especially virtual environments in which prosperous businesses and communities work confidently. Reputation plays great importance in enabling trusted business interactions. Trust and reputation are indispensable and pivotal for regulation based on quality. Normally trust and reputation are used successfully in building architectures for quality management.

References

1. Marin, C.A., Stalker, I., Mehandjiev, N.: Business Ecosystem Modelling: Combining Natural Ecosystems and Multi-Agent Systems. In: Klusch, M., Hindriks, K.V., Papazoglou, M.P., Sterling, L. (eds.) CIA 2007. LNCS, vol. 4676, pp. 181–195. Springer, Heidelberg (2007)
2. Vuori, E.K.: Knowledge-intensive service organizations as agents in a business ecosystem. In: International Conference on Services Systems and Services Management, pp. 908–912. IEEE, Tampere (2005)
3. Muntaner-Perich, E., de la Rosa Esteva, J.L.: Using Dynamic Electronic Institutions to Enable Digital Business Ecosystems. In: Noriega, P., Vázquez-Salceda, J., Boella, G., Boissier, O., Dignum, V., Fornara, N., Matson, E. (eds.) COIN 2006. LNCS, vol. 4386, pp. 259–273. Springer, Heidelberg (2007)
4. Kraus, S., Shehory, O., Taase, G.: Coalition formation with uncertain heterogeneous information. In: AAMAS 2003, pp. 1–8. ACM, New York (2003)
5. Klusch, M., Gerber, A.: Dynamic Coalition Formation among Rational Agents. IEEE Intelligent Systems 17(3), 42–47 (2002)
6. Muntaner-Perich, E., de la Rosa Esteva, J.L.: Towards a Formalisation of Dynamic Electronic Institutions. In: Sichman, J.S., Padget, J., Ossowski, S., Noriega, P. (eds.) COIN 2007. LNCS, vol. 4870, pp. 97–109. Springer, Heidelberg (2008)
7. Tsvetovat, M., Sycara, K., Chen, Y., Ying, J.: Customer Coalitions in Electronic Marketplace. In: Proceedings of the Fourth International Conference on Autonomous Agents, pp. 263–264. ACM, New York (2000)
8. Westwood, K., Allan, V.H.: Heuristics for Co-opetition in Agent Coalition Formation. In: SCAI 2006, pp. 143–150 (2006)
9. Ghedira, C., Maamar, Z., Benslimane, D.: On Composing Web Services for Coalition Operations - Concepts and Operations. International Journal Information & Security. Special issue on Architectures for Coalition Operations 16, 79–92 (2005)
10. Chang, E., Dillon, T., Hussain, F.K.: Trust and Reputation for Service-Oriented Environments: Technologies for building Business Intelligence and consumer confidence. John Wiley & Sons, Chichester (2006)
11. Berardi, D., Calvanese, D., De Giacomo, G., Lenzerini, M., Mecella, M.: A Foundational Vision for e-Services. In: Ubiquitous Mobile Information and collaboration Systems Workshop (2003)
12. Scully, T., Madden, M.G., Lyons, G.: Coalition Calculation in a Dynamic Agent Environment. In: Twenty-first international conference on Machine learning (ICML 2004), p. 93. ACM, New York (2004)

Future Information Technology for the Health Sector
- A Delphi Study of the Research Project FAZIT -

Kerstin Cuhls[1], Simone Kimpeler[1], and Felix Jansen[2]

[1] Fraunhofer ISI
[2] MFG Stiftung Baden-Württemberg

Abstract. Information technology in the health sector will continue to be an important topic in the oncoming years. This offers interfaces for new market potential for IT companies. However, which information technologies bring about change? This was the initial question for a Delphi study in the context of the research project FAZIT. In order to find answers to this question, information technological developments were identified, which could become relevant during the next 20 years.

1 Information Technology for the Health Sector: Which Are the Technical Challenges?

Information Technology will (continuously) alter the Health Sector strongly in the coming years. But in how far? And which technologies will be those evoking these alterations? This was the core question for the study about Information Technology in the Health Sector at hand. Where and in which way can Information and Communication Technology (ICT) contribute to improving or economising the Health System? What is, though technically possible, not desirable?

By means of this Delphi Study about "Information Technology in the Health Sector" it has been analysed which specific contributions by Information Technology at all can improve the Health Sector, when they can be realised and which obstacles need to be overcome on the way towards realisation.

In order to answer these questions, the IT developments relevant for the next 20 years were identified and phrased in the form of theses. A two-step survey was conducted, during which experts evaluated the theses according to their importance, feasibility and desirability. Even during the time of identifying the theses it became apparent, that although technology an indeed contribute to the sustainability of the Health System, the success of the technological innovations is strongly related to such factors as costs, acceptance, regulation or organisational alterations.

The study at hand represents a part of a larger project called FAZIT (research project for current and future information and media technology and its use in Baden-Wuerttemberg), which is facilitated by the state of Baden Wuerttemberg. Since the beginning of 2005, in the context of FAZIT the demand and applicability for innovative information and media technology has been analysed and longterm drivers, which lead to new market opportunities for ICT in Baden-Wuerttemberg, are

M. Ulieru, P. Palensky, and R. Doursat (Eds.): IT Revolutions 2008, LNICST 11, pp.122–139, 2009.

identified. The project is supported by the MFG Foundation Baden-Wuerttemberg (MFG Stiftung), the Centre for European Economic Research (ZEW) and the Fraunhofer Institute for Systems and Innovation Research (Fraunhofer ISI) partner it.

A multi-step foresight process in the framework of the project, conducted by the Fraunhofer ISI, identifies the research and development units which are important for the innovation potential of the state of Baden-Wurttemberg. As this phenomenon reflects social as well as economical trends, a combination of foresight methods is applied. Social and technological developments are screened in three Delphi studies and are evaluated by experts regarding specific criteria such as their importance or their realisation time. The results enter a scenario process, in order to assess the future viability of Baden-Wurttemberg in the ICT sector and in order to delineate selected market segments.

The report at hand describes the second Delphi study of the project including the conduct of the survey and its method. Subsequently this text gives an overview of the results of all theses. The report is concluded with a short glimpse into the future. Detailed evaluations for each thesis and the results are described in the extensive research report which can be download free of charge on the website www.fazit-research.org.

1.1 Why and How to Look into the Future?

The FAZIT research project is scientifically embedded in the research approach of regional innovation systems, which understands innovation as an evolutionary and cumulative process that delivers feedback. Innovations can only be realised in economic and social interaction of various regional players and result in technological, organisational and social changes (Koschatzky, 2001). This, on the one hand, emphasises the social angle of innovation in the sense of a collective learning process and, on the other hand, the high relevance of involving all players of the region. The approach of regional innovation systems leads to the conclusion, that the future viability of Baden-Wurttemberg depends especially on the fact, how successfully knowledge is generated, newly associated and implemented in products. The key factor to successful innovation in the regional context is an institutionalised network between enterprises, universities and organisations as well as the social structure of innovations in the field of Technology Push and Market Pull (Leydesdorff, 2005).

Especially long-term research, long-range technological development or the influence of social mega trends brings about new products and profitable markets. In order to be able to outline long-term perspectives and design stable future prospects and not merely gain shortterm vantage points, FAZIT combines various methods of future research. It is the aim, to embed the results globally, however, to focus on the location Baden-Wurttemberg and to point out specific local respectively regional challenges, in order to enable new strategies or to adjust and re-align existing strategies. The so-called "regional foresight" approach thus facilitates strategic decision-making on behalf of all players in the innovation system. The players are involved in the process of future research.

Future prospects should be linked to today's decision processes. This way, they can, today, facilitate appropriate decisions and trigger acts, which are aligned according to a common future.

When designing future prospects one is very well aware of the fact, that the future is not predictable. However, there are certain developments, which can in fact be taken into consideration, especially in the fields of science and technology. When applying foresight methods, experts are primarily interested in those things, which are, above all, on the agenda. For this reason, foresight is a systematic glance into the future in terms of economy, science and society with the aim of identifying those fields of strategic research and new technology, which will most probably have a strong impact on economy and the well-being of human beings (see Martin, 1995 and Cuhls, 2003, respectively).

1.2 What Is the Delphi Method?

The Delphi Method was developed in the 1950s by the Rand Corporation, Santa Monica, California, as a method of "operations research" (a type of system research, which uses statistics, mathematical models etc. for decision-making) for military research. In Japan, it has been applied in national, civil context ever since the beginning seventies and has become an element of foresight processes throughout Europe since the beginning nineties (compare Blind/Cuhls/Grupp, 1999; 2002). The Fraunhofer Institute for Systems and Innovation Research conducted some of the first German national Delphi surveys for the Federal Ministry of Research and Technology (today Federal Ministry of Education and Research, Bundesministerium für Bildung und Forschung – BMBF). Ever since then, the method has been internationally refined together with a Japanese partner (Cuhls/Kuwahara, 1994; Cuhls/Breiner/Grupp, 1995). Thanks to new possibilities of electronic surveying, especially in the European and Asian context, the Delphi Method has gained more and more popularity (e.g. EUFORIA, FISTERA, Delphi of the Millennium Project; NISTEP, 2005; MOST, 2003 and 2005).

Delphi processes are generally surveys conducted in two or more rounds or "waves". As from the second round, feedback is given on the first round. The topics to be evaluated are generated through different sources, desk research or group processes. The interviewees are mostly professionals, often decision-makers from sectors such as the economy, research, but also associations or other organisations. During the second round, each interviewee can consider whether he or she will take the given evaluations of all questioned experts into account and be influenced by them (Häder/Häder, 2000).

2   Overview of Results

2.1 When Are the Theses Considered Feasible?

The timeframe for realisation was not polled exactly, moreover in five-year steps, since nobody knows exactly what lies in the future. This is shown with a median and the quartiles Q1 (25 percent step) and Q2 (75 percent step). Exactly half of the participants give answers that are located between the ratios of Q1 and Q2. One quarter gives evaluations that are ranged below this timeframe, one quarters' evaluations range above it. This way a range of opinions can be described. If the lower (Q1) and upper quartile (Q2) are very close to each other, half of the participants (generally even the majority) are consistent in their evaluation. If they are further or far separated, there is no or hardly any consensus or there may be large precariousness about the time of realisation. Table 1 shows when participants expect the realisation, for all theses. The theses are sorted according to the median, meaning the 50 percent mark, ranging from early realisations to later ones. The median can be calculated by help of a complex formula (compare Cuhls/Kuwahara, 1994; BMFT, 1993). There is an easier way to calculate it, as well: The answers are sorted according to the five-year steps from 2006 to 2010, from 2011 to 2015, from 2016 to 2020, from 2021 to 2025, from 2026 to 2030 and "later". Afterwards, the number of answers is counted, divided by two and claried as to where the answer is located (assuming an even distribution within each marking box). Example: 100 persons answered, each box received 20 marks, and nobody marked "later". This means, the 50th respectively 51st person's answer is located in the exact middle of the box ranging from 2016 to 2020, the year being 2018. This is the median.

Table 1. Realisation of all 36 theses sorted by year of realisation (early to late)

Theses	Year of realisation (50 percentage point)
Expert systems and databases, which monitor customised medications for individual patients with respect to undesired medication interactions and recommendations for a pharmaceutical therapy with reduced adverse reactions and side effects, are tested in pilot experiments.	2010
Patients in hospitals are directed by an EDP-supported planning system, so that waiting periods, e.g. at admission, diagnostic procedures (X-ray, CT, endoscopy, etc.), operation are minimised and at the same time the overall efficiency of hospital facilities is enhanced.	2010
Regional microwave hyperthermia can be ideally planned with a computer simulation of the biothermal conduction.	2012
A computerised system exists, which allows practice-based physicians to access all information at hand about the patient (cryptographically secured) via a terminal of their choice during house calls.	2012
Virtual reality is a standard in training of medical staff (e.g. virtual surgery, practising of minimally invasive interventions, endoscopy, rescue practices, patient interviews etc.).	2012
A non-invasive long-term blood pressure sensor has been developed.	2012

Table 1. (*continued*)

Documentation tasks in hospitals are routinely performed via voice entry.	2013
Telemonitoring, i.e. close-meshed monitoring of patients (at risk), evaluation of the generated information in and by medical facilities and, if necessary, alerting the treating physician, has become a standard.	2013
Wireless rechargeable implanted defibrillators are used, which convey their measured data to a control unit, which then conveys its data to a service centre for a check up and for an emergency report, if necessary.	2013
Ambient Intelligence in a house allows monitoring of patients at home (via camera, thinking carpet, furniture equipped with sensors, immobility sensors), reporting irregular features to an emergency call centre.	2013
An implantable data carrier has been developed, storing all data of a patient necessary for treatment and administration.	2013
Labs-on-Chips are broadly applied for "point of care" diagnoses of clinically relevant parameters such as proteins, antibodies, hormones, bilirubine, cholesterol, urea as well as enzymes in blood and urine.	2013
Computer-supported planning of biologically adaptive resonance therapy (ART), which allows an individual adaptation of the therapy to heterogeneous tissue, is possible.	2013
Voice recognition and correct relation of a voice to the person speaking is so accurate, that surgeons are able to navigate instruments through voice commands and are thus effectively relieved.	2013
Electrodes in the brain detect a beginning epileptic seizure and prevent it through specific electrical stimulation patterns.	2013
Expert systems are routinely appointed to recommend specific advice for diagnoses and therapies to the healthcare staff.	2014
A wireless label system (RFID) is introduced to common households, allowing patients who easily and often forget things (due to dementia, Alzheimer's disease etc.) to find anything and be attentive to things of importance.	2014
Technologies are applied in research, which allow forecasts on biological activity of proteins and their functional domains via information as to their spatial configuration.	2014
Interactive electronic logopaedics trainers are a standard.	2014

Table 1. (*continued*)

Fully functional robot systems have been developed and tested for transdermal intervention (e.g. biopsy robots).	2014
Valid diagnostic test procedures based on functional Magnetic Resonance Imaging (MRI) are clinically used for diagnoses with mental diseases (e.g. manic-depressive diseases) and diseases of the central nervous system (e.g. Alzheimer's disease).	2014
Routine whole-body scanning with functional imaging is a standard procedure after accidents.	2014
Protein-chips for "Point of Care" diagnostics have been developed and tested.	2014
Histological diagnosis of tissue in vivo is possible with the help of spectroscopic, microscopic laser scanning methods.	2014
Clinically applicable systems consisting of implantable glucoses sensors, actuators and insulin reservoirs as well as corresponding control software have been developed, allowing an optimum fine-tuning of diabetes patients.	2014
In emergency cases, in order to be able to identify a person very soon after an accident, a quick genetic test is completed and the data is matched with a profile database.	2016
Due to IT approaches (simulations, virtual animal models), 80% of all animal testing in medical and pharmaceutical research becomes redundant.	2016
Methods for quick analysis of the genome, e.g. DNA Chips, high-speed sequencing or genetic mapping are applied in healthcare routine.	2017
Vital parameters (blood pressure, blood levels, antibodies, hormones) can be deciphered via implanted chips.	2017
Blind persons can orient themselves within a room with a retina implant.	2018
Standardisation and processing of the large mass of data delivered through proteomics has developed a predictive and integrative biology, consisting of techniques for visualising results, automatic matching with other genome-comprising data records as well as the integration of additional "-omics" (genomics etc.) approaches.	2018
Many hospitals employ robots for difficult and standard procedures in nursing (e.g. putting someone into another bed, changing of bedclothes) in order to relieve the nursing staff and enable them to have more time for personal attentiveness towards the patients.	2018

Table 1. (*continued*)

Retina implants improve dramatically and thus become ready for use through combination of functional and morphological data, the evaluation of the data by expert systems and the cross linking of the various systems.	2019
Surgeries within the body, which are conducted by a remote-controlled micromachine, equipped with sensors and actuators, are possible.	2019
Entire artificial kidneys have been developed.	2022
An artificial heart and lung implant receives marketing approval.	later

The scale of evaluations in the questionnaire ranges up to the year 2030, which means that it is not possible to give an exact estimate for evaluations dated after this, the table simply lists "later". The estimate "never" was calculated by percent and only mentioned when the ratio is very high.

The realisations expected later are either very specific, extremely controversial, as "operations conducted via micro machines", respectively "nursing robots", or they pose technical and technological challenges, like the two theses concerning artificial organs "fully artificial kidneys are developed" and "an artificial heart and lung implant receives market approval". Correspondingly, such theses can only become reality at an earlier stage if there are measures taken into consideration, which promote them.

All theses are considered feasible, only a small minority marked "never" realisable for the categories – which will be explained at a later stage – "desirability" and "importance" of the theses considered most controversial. These are the five theses:

1. Many hospitals employ robots for difficult and standard procedures in nursing (e.g. putting someone into another bed, changing of bedclothes) in order to relieve the nursing staff and enable them to have more time for personal attentiveness towards the patients. (nearly 24 percent say "never")
2. Due to IT approaches (simulations, virtual animal models), 80% of all animal testing in medical and pharmaceutical research becomes redundant. (12 percent say "never")
3. An implantable data carrier has been developed, storing all data of a patient necessary for treatment and administration. (12 percent say "never")
4. Voice recognition and correct relation of a voice to the person speaking is so accurate, that surgeons are able to navigate instruments through voice demands and are thus effectively relieved. (10 percent say "never")
5. In emergency cases, in order to be able to identify a person very soon after an accident, a quick genetic test is completed and the data is matched with a profile database. (nearly 10 percent say "never")

2.2 Nearly All Selected Theses Are Desirable

In this Delphi study the desirability of the theses was directly determined. The question posed was: Do you personally find the realisation of this thesis altogether desirable? Nearly all topics open for discussion were considered desirable (mostly more than 90 percent). There even are theses, which gained 100 percent approval, they are:

- Entire artificial kidneys have been developed.
- Virtual reality is a standard in training of medical staff (e.g. virtual surgery, practising of minimally invasive interventions, endoscopy, rescue practices, patient interviews etc.).
- Blind persons can orient themselves within a room with retina implants.
- Computer-supported planning of biologically adaptive resonance therapy (ART), which allows an individual adaptation of the therapy to heterogeneous tissue, is possible.

Only four theses are controversial as to their desirability, i.e. their ratio for "desirability: yes" is far lower than the mean ratio, though still fairly high:

- An implantable data carrier has been developed, storing all data of a patient necessary for treatment and administration. (64 percent say "no")
- Many hospitals employ robots for difficult and standard procedures in nursing (e.g. putting someone into another bed, changing of bedclothes) in order to relieve the nursing staff and enable them to have more time for personal attentiveness towards the patients. (54 percent say "no")
- In emergency cases, in order to be able to identify a person very soon after an accident, a quick genetic test is completed and the data is matched with a profile database. (20 percent say "no")
- Methods for quick analysis of the genome, e.g. DNA Chips, high-speed sequencing or genetic mapping are applied in healthcare routine. (14 percent say "no", 14 percent say "do not know")

Two of these theses, which are not necessarily desirable, show the highest ratios for "never realisable". Perhaps they will not be realised or only realised at a later stage, since, in their contemporary form, they do not belong to the desirable theses. Especially controversial, for many years now, are the "robots in nursing" (see also BMFT, 1993 or Cuhls/Blind/Grupp, 1998). Commentaries of participants indicate that they will become an inevitable necessity regarding the "healthcare crisis" and demographic change, which will probably aggravate problems in nursing and healthcare. Prototypes of such robots in fact, already exist (e.g. Care- O-Bot at the Fraunhofer Institute for Production Technology and Automation, IPA, 2006). Alternative solutions to robotised nursing need to be found on other levels apart from the purely technological one (attractiveness of nursing occupations, salaries etc.) The technology as such will surely be applied in other service areas, too – or even earlier.

2.3 Quality of Healthcare and Importance of Higher Quality of Life

The next question referred to the importance of the thesis, not so much overall, moreover on a cost reduction basis, technological progress, better healthcare provision, quality of healthcare, quality of living, environmental protection, sustainability or others. Evaluations by Delphi experts show that the importance of the averagely most often mentioned theses is laid upon a better quality of healthcare and higher quality of life for human beings. However, both were only marked by around half of the participants. The values are not very high, suggesting the ranking of the theses' mean importance not very high, either. Since multiple marking was allowed, the naming of topics is broadly distributed. In this manner, cost reduction, better healthcare provision and technological progress were marked as well, however, values varied between respective theses, which makes the average depicted in illustration 2 less convincing. As cost reduction was marked for one thesis and other categories more often for others, the marks spread out in the percentage analysis.

Illustration 1. What are the topics important for?

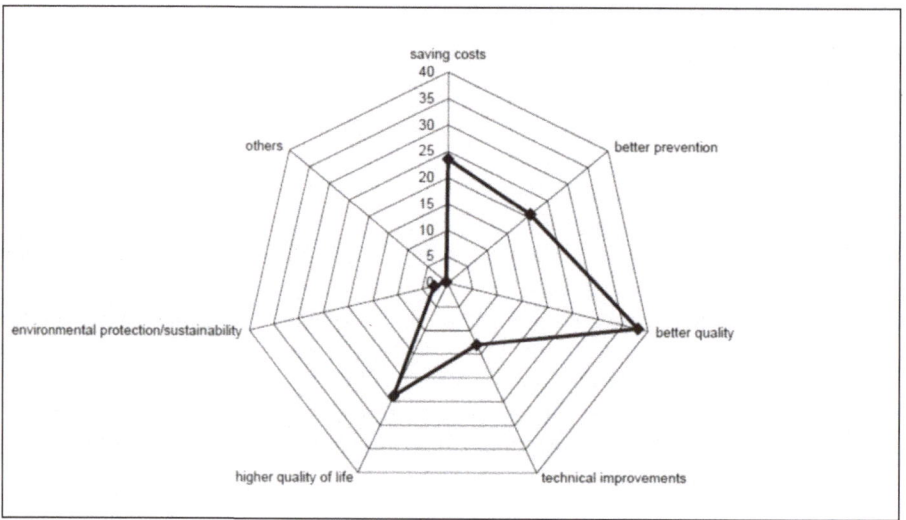

The theses are not of importance for technological progress itself or environmental protection or sustainability. The latter is not surprising, since the topics dealt with concern application-related healthcare. The only thesis playing a major part in terms of environmental protection and sustainability is: "Due to IT approaches (simulations, virtual animal models), 80% of all animal testing in medical and pharmaceutical research becomes redundant.".

However, what can be shown is the fact that even though the development of new information technology will be important for applications, it will not serve as a driver for technological progress as such. The only topic considered important for technological process by more than 80 percent of the experts is: "Standardisation and processing of the large mass of data delivered through proteomics has developed a predictive and integrative biology, consisting of techniques for visualising results, automatic matching with other genomecomprising data records as well as the integration of additional "-omics" (genomics etc.) approaches.". Does information technological progress only provide such few impulses to the healthcare sector?

As a selection the following theses respectively are important for a better quality of healthcare, more than 90 percent of the participants considered them important:

- Computer-supported planning of biologically adaptive resonance therapy (ART), which allows an individual adaptation of the therapy to heterogeneous tissue, is possible.
- A computerised system exists, which allows practice-based physicians to access all information at hand about the patient (cryptographically secured) via a terminal of their choice during house calls.
- Patients in hospitals are directed by an EDP-supported planning system, so that waiting periods, e.g. at admission, diagnostic procedures (X-ray, CT, endoscopy, etc.), operation are minimised and at the same time the overall efficiency of hospital facilities is enhanced.
- Telemonitoring, i.e. close-meshed monitoring of patients (at risk), evaluation of the generated information in and by medical facilities and, if necessary, alerting the treating physician, has become a standard.
- Histological diagnosis of tissue in vivo is possible with the help of spectroscopic, microscopic laser scanning methods.
- Expert systems and databases, which monitor customised medications for individual patients with respect to undesired medication interactions and recommendations for a pharmaceutical therapy with reduced adverse reactions and side effects, are tested in pilot experiments.
- Surgeries within the body, which are conducted by a remote-controlled micromachine, equipped with sensors and actuators, are possible.
- Regional microwave hyperthermia can be ideally planned with a computer simulation of the biothermal conduction.
- Virtual reality is a standard in training of medical staff (e.g. virtual surgery, practising of minimally invasive interventions, endoscopy, rescue practices, patient interviews etc.).

For better healthcare provision the following two theses were considered to be the most important by more than 80 percent of the Delphi experts:

- Labs-on-Chips are broadly applied for "point of care" diagnoses of clinically relevant parameters such as proteins, antibodies, hormones, bilirubine, cholesterol, urea as well as enzymes in blood and urine.
- A non-invasive long-term blood pressure sensor has been developed.

Six of the theses were considered to be especially important with reference to cost reduction; predominantly, with a share of more than 85 percent of the participants, naming systems with software solutions for data processing (the first three theses). The other three theses were named by more than 80 percent of the participants.

- Patients in hospitals are directed by an EDP-supported planning system, so that waiting periods, e.g. at admission, diagnostic procedures (X-ray, CT, endoscopy, etc.), operation are minimised and at the same time the overall efficiency of hospital facilities is enhanced.
- Documentation tasks in hospitals are routinely performed via voice entry.
- A computerised system exists, which allows practice-based physicians to access all information at hand about the patient (cryptographically secured) via a terminal of their choice during house calls.

- Virtual reality is a standard in training of medical staff (e.g. virtual surgery, practising of minimally invasive interventions, endoscopy, rescue practices, patient interviews etc.).
- Labs-on-Chips are broadly applied for "point of care" diagnoses of clinically relevant parameters such as proteins, antibodies, hormones, bilirubine, cholesterol, urea as well as enzymes in blood and urine.
- Due to IT approaches (simulations, virtual animal models), 80% of all animal testing in medical and pharmaceutical research becomes redundant.

2.4 Applicability in Other Areas

The named information technology theses are not only applicable in the health sector; many of the technologies can be applied to other areas. The choice offered the sectors product development, management, industry and production, logistics, sales, environmental management, traffic and mobility and "others" (see illustration 2). Surprisingly, for each thesis different sectors were named, which means each of the technology approaches offered for discussion is applicable in another area. This demonstrates the objective of demonstrating how information technological approaches have been selected, which will not only concern the health sector, but be of interest to other "markets" as well.

The application named alternatively in most cases was product development. Even though the mean evaluation by the Delphi experts resembles 28 namings the ratio can be considered high, as so-called technological approaches were not estimated to be that diverse. None of the technologies in the theses is considered applicable for "none" of the named areas. "Other" areas than those offered could be named and phrased, this, however only took place rarely (average one mark), this is why they are not represented in illustration 2.

Illustration 2. Application of the technology of the theses in other areas

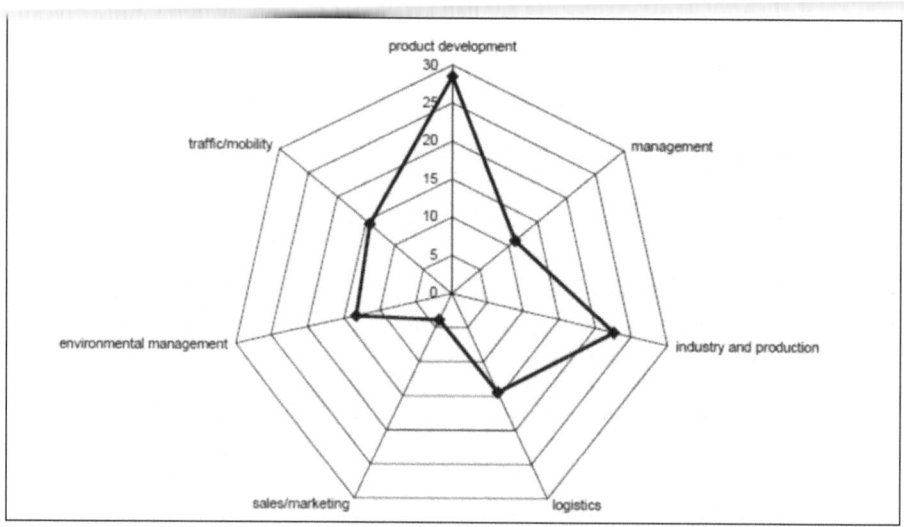

2.5   Technical Problems Are the Main Obstacle for Realisation

Even though the participants in the Delphi study rated technical problems as the largest obstacle (see illustration 3, based on the number of namings) they believe problems are solvable. This precisely is the reason why the topics were included in the Delphi study: They should represent a technical challenge; on the other hand they need to seem realistic. Thus, the selected topics seem to be the "right" ones, particularly since they receive relatively high ratings in the evaluations on importance (see above).

Illustration 3. Obstacles for realisation

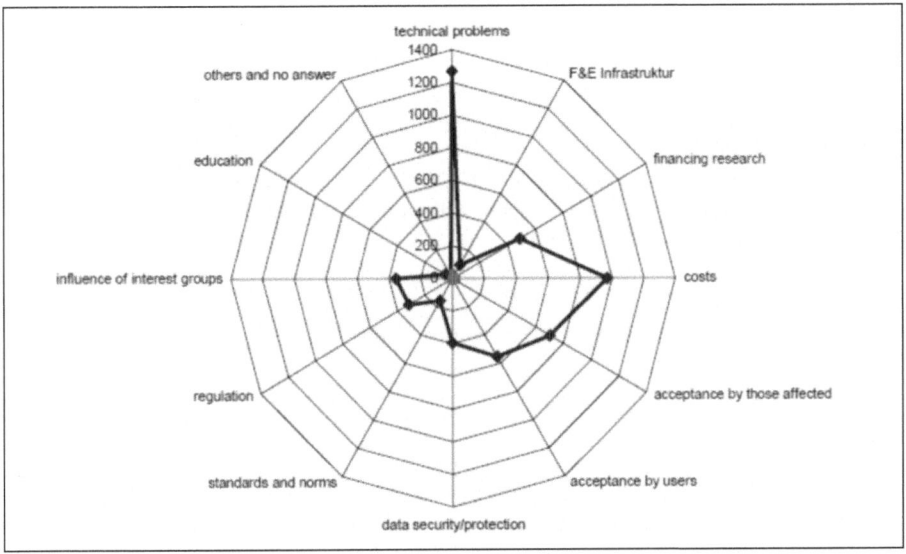

Note: All namings/markings were counted and divided by the number of theses in order to obtain a mean value.Multiple namings were possible.

35 participants among 86 participants in the second round marked "technical obstacles" on average. The 10 most-marked theses are presented in table 2. Respectively more than 90 percent of the participants marked a cross here.

The analysis of the "obstacles for realisation" shows, that, apart from technical problems, realisation of individual theses could fail due to the high costs. Overall costs were not mentioned often, but came second in the field of realisation problems – even if cost reduction may serve as a reason for realisation of the theses. In singular cases, this contradiction is rare to be found and then generally refers to topics, which are connected to high investment expenditure and may only lead to long-term cost reduction and saving.

Table 2. Top 10 theses posing the largest technical problems

Theses	Namings in percent
Retina implants improve dramatically and thus become ready for use through combination of functional and morphological data, the evaluation of the data by expert systems and the cross linking of the various systems.	100
Entire artificial kidneys have been developed.	95,5
Computer-supported planning of biologically adaptive resonance therapy (ART), which allows an individual adaptation of the therapy to heterogeneous tissue, is possible.	94,6
Surgeries within the body, which are conducted by a remote-controlled micromachine, equipped with sensors and actuators, are possible.	94,5
Clinically applicable systems consisting of implantable glucoses sensors, actuators and insulin reservoirs as well as corresponding control software have been developed, allowing an optimum fine-tuning of diabetes patients.	93,8
An artificial heart and lung implant receives marketing approval.	93,2
Blind persons can orient themselves within a room with a retina implant.	92,6
A non-invasive long-term blood pressure sensor has been developed.	91,8
Voice recognition and correct relation of a voice to the person speaking is so accurate, that surgeons are able to navigate instruments through voice commands and are thus effectively relieved.	91,4
Methods for quick analysis of the genome, e.g. DNA Chips, high-speed sequencing or genetic mapping are applied in healthcare routine.	90,7

In detail more than 80 percent of the participants marked costs as an obstacle for the following theses:

- Routine whole-body scanning with functional imaging is a standard procedure after accidents.
- Ambient Intelligence in a house allows monitoring of patients at home (via camera, thinking carpet, furniture equipped with sensors, immobility sensors), reporting irregular features to an emergency call centre.
- Telemonitoring, i.e. close-meshed monitoring of patients (at risk), evaluation of the generated information in and by medical facilities and, if necessary, alerting the treating physician, has become a standard.

- Patients in hospitals are directed by an EDP-supported planning system, so that waiting periods, e.g. at admission, diagnostic procedures (X-ray, CT, endoscopy, etc.), operation are minimised and at the same time the overall efficiency of hospital facilities is enhanced.
- Valid diagnostic test procedures based on functional Magnetic Resonance Imaging (MRI) are clinically used for diagnoses with mental diseases (e.g. manic-depressive diseases) and diseases of the central nervous system (e.g. Alzheimer's disease).

Obstacles which were named as well - even though not as frequently – are research funding as well as acceptance by those concerned and users. Data protection is a factor not to be underestimated in rare cases, either. However, it is interesting to note that research and development infrastructure, standards and norms, education and further education and other obstacles are not critical for the experts.

3 A Short Glance at the Future: Where Do New Markets Develop?

Human beings tend to orient themselves by conspicuities and ignore the rather normal "boringly evolving topics". For this reason, the reference may be permitted to state that the most attention attracting topics and in certain cases controversial theses of the Delphi study at hand may be of the highest media and public interest. However, this does not always apply to interest regarding market developments, as the mentioned technologies often involve risks for developers and manufacturers. For this reason, as a user of the data published in the scope of the study at hand, you need to specify between the different criteria respectively questions and use them as a framework for an analysis of your own, instead of simply adopting those topics which easily catch the eye on account of certain results.

For instance, it is assumed that "robots in nursing" and "biopsy robots" – each of the topics being considered desirable only to a certain degree and frequently discussed in commentaries – will become reality and thus appear on the market later than the mean subjects of other theses.

Theses, which are of interest to the industry, e.g. technologies that can be applied and transferred to other areas, should be closely watched. They represent the most interesting and important topics of the study, in spite of receiving rather moderate estimates in terms of realisation dates – most often they are "only" located in the upper respectively early midfield of scheduled market appearance.

For new markets in Baden-Wurttemberg topics such as proteomics, telemonitoring, voice entry for documentation tasks or data access from everywhere, which do not sound spectacular but require a high level of knowledge and know-how, are especially interesting. Of course, to a certain and large degree, they coincide with those topics discussed on a national level and which have, partly, been adopted by the Federal Government's high-tech-strategy (BMBF, 2006 and 2007). Only these technology approaches will lead to broad market application in other sectors. What goes without saying is the fact that for this Delphi study, all theses were selected with regard to their importance and impact on future development, some of them most probably developing in a more dynamic way than others.

One of the topics with an especially dynamic development is the retina implant with two theses. Both are considered very desirable, as they are extremely important for the quality of life of visually impaired and blind persons. That is why efforts already have been made in this area, in spite of high obstacles and costs, which will prove the estimates by Delphi experts to be wrong. Their estimates are very pessimistic and broadly distributed in terms of realisation times (median 2018 and 2020, respectively). Sometimes it is better if prognoses regarding time scopes are not fulfilled. Labelled as autodestructing prophecies (prognoses estimating "never" or wrong timeframes on account of decisions based on today's knowledge), the complete retina implant topic could become one of those topics not "being realised" and thus trigger a possibly even earlier apparition than estimated here and now. Before, however, many technical obstacles need to be overcome.

Many of the selected technical topics are not restricted to the health sector, but equally relevant for other areas of application. In particular product development was frequently named, but also industry and production as well as logistics. It was not expected that the Delphi participants marked so many areas of application. Sometimes, interpretation is difficult (What is meant by "product development"?), if commentaries do not supply further references. Which "new" products are meant? In the most cases, this was not elaborated upon. The signal, however, indicates that even though all selected topics regard information technology as well as health as such, the supporting technology is not confined to this sector. All theses referring to voice entry, for instance, could be applied in a variety of other areas, as well. Perhaps even at an earlier stage, since the accuracy, which needs to be guaranteed in an operating theatre should be reliably close to 100 percent. For other applications, such as secretarial tasks, this is not mandatory.

Noticeable but not surprising is the fact that topics which literally "get under the skin" are very controversial. In the Delphi study at hand this concerns all theses regarding implantation as well as transdermal intervention.

Generally speaking all these theses are considered desirable, however, they do not attain evaluation ranks as positive as others. Apart from technical problems issues regarding data protection and data security or acceptance by those concerned are named. These topics also raise ethical concerns and commentaries refer to rejection based on more emotional, rather than rational grounds.

Nearly all of the theses are considered realisable, however, obstacles must not be underestimated. The following theses, for instance, could fall through on account of related costs, since the item "costs" as an obstacle is often marked:

- Routine whole-body scanning with functional imaging is a standard procedure after accidents.
- Ambient Intelligence in a house allows monitoring of patients at home (via camera, thinking carpet, furniture equipped with sensors, immobility sensors), reporting irregular features to an emergency call centre.
- Telemonitoring, i.e. close-meshed monitoring of patients (at risk), evaluation of the generated information in and by medical facilities and, if necessary, alerting the treating physician, has become a standard.
- Patients in hospitals are directed by an EDP-supported planning system, so that waiting periods, e.g. at admission, diagnostic procedures (X-ray, CT,

endoscopy, etc.), operation are minimised and at the same time the overall efficiency of hospital facilities is enhanced.

- Valid diagnostic test procedures based on functional Magnetic Resonance Imaging (MRI) are clinically used for diagnoses with mental diseases (e.g. manic-depressive diseases) and diseases of the central nervous system (e.g. Alzheimer's disease).

Some of the above listed theses are considered to be very important, e.g. the topic "ambient intelligence", which will become important for an improvement of quality of life and also the technology may be found applicable in other areas. None of the theses is considered to be unrealisable, and if appropriately planned (cost reduction, large number of applications respectively more use, simplification of technology etc.) the problems regarding costs can be solved.

Technical issues pose more critical problems named as obstacles for certain theses by all or nearly all participants. Worth mentioning here, again, retina implants, artificial organs, micro machines and implantable minimal systems, but also radiotherapy planning, voice recognition, analysis of the genome in healthcare routine and a "noninvasive long-term blood pressure sensor".

New markets are expected in the areas of voice recognition, virtual reality and simulations, database approaches, sensory development, radio frequency identification (RFID) or new management and planning systems, either because they are generally considered important or because they reduce costs or the technology behind is so comprehensive that is not only applicable in the health sector, but also in other areas or sectors. In particular this criterion is valid for the following theses:

- Expert systems and databases, which monitor customised medications for individual patients with respect to undesired medication interactions and recommendations for a pharmaceutical therapy with reduced adverse reactions and side effects, are tested in pilot experiments.
- Patients in hospitals are directed by an EDP-supported planning system, so that waiting periods, e.g. at admission, diagnostic procedures (X-ray, CT, endoscopy, etc.), operation are minimised and at the same time the overall efficiency of hospital facilities is enhanced.
- A computerised system exists, which allows practice-based physicians to access all information at hand about the patient (cryptographically secured) via a terminal of their choice during house calls.
- Regional microwave hyperthermia can be ideally planned with a computer simulation of the biothermal conduction.
- Virtual reality is a standard in training of medical staff (e.g. virtual surgery, practising of minimally invasive interventions, endoscopy, rescue practices, patient interviews etc.).
- A non-invasive long-term blood pressure sensor has been developed.
- Documentation tasks in hospitals are routinely performed via voice entry.
- Telemonitoring, i.e. close-meshed monitoring of patients (at risk), evaluation of the generated information in and by medical facilities and, if necessary, alerting the treating physician, has become a standard.

- Labs-on-Chips are broadly applied for "point of care" diagnoses of clinically relevant parameters such as proteins, antibodies, hormones, bilirubine, cholesterol, urea as well as enzymes in blood and urine.
- Computer-supported planning of biologically adaptive resonance therapy (ART), which allows an individual adaptation of the therapy to heterogeneous tissue, is possible.
- Expert systems are routinely appointed to recommend specific advice for diagnoses and therapies to the healthcare staff.
- A wireless label system (RFID) is introduced to common households, allowing patients who easily and often forget things (due to dementia, Alzheimer's disease etc.) to find anything and be attentive to things of importance.
- Wireless rechargeable implanted defibrillators are used, which convey their measured data to a control unit, which then conveys its data to a service centre for a check up and for an emergency report, if necessary.
- Interactive electronic logopaedics are a standard.
- Protein-chips for "Point of Care" diagnostics have been developed and tested.
- Technologies are applied in research, which allow forecasts on biological activity of proteins and their functional domains via information as to their spatial configuration.
- Histological diagnosis of tissue in vivo is possible with the help of spectroscopic, microscopic laser scanning methods.
- Clinically applicable systems consisting of implantable glucoses sensors, actuators and insulin reservoirs as well as corresponding control software have been developed, allowing an optimum fine-tuning of diabetes patients.

Of all the above mentioned, the following two theses are considered most important for better prevention. They were named by more than 80 percent of the Delphi experts:

- Labs-on-Chips are broadly applied for "point of care" diagnoses of clinically relevant parameters such as proteins, antibodies, hormones, bilirubine, cholesterol, urea as well as enzymes in blood and urine.
- A non-invasive long-term blood pressure sensor has been developed.

These technologies – which are, sometimes, not even spectacular – and their application may open doors to new markets, in Baden-Wuerttemberg as well as in other places. The chances of their realisation during the next 10 to 15 years are not bad. Although technical obstacles remain and must not be disregarded they seem surmountable. As described in chapter 4 and recorded in patent applications, some enterprises already seem to be waiting in their starting blocks.

References

Blind, K., Cuhls, K., Grupp, H.: Current Foresight Activities in Central Europe. Technological Forecasting and Social Change, Special Issue on National Foresight Projects 60(1), S. 15–37 (1999)

Bundesministerium für Forschung und Technologie (BMBF, Hg.): Aktionsplan Medizintechnik 2007-2008. Part of the Hightech-Strategy of the Federal Government, Bonn and Berlin(2007), http://www.gesundheitsforschung-bmbf.de (accessed Feburary 21,2007)

Bundesministerium für Forschung und Technologie (BMBF, Hg.): Ideen zünden. Die Hightech-Strategie für Deutschland, Bonn and Berlin (2006), http://www.bmbf.de (accessed Feburary 21, 2007)

Bundesministerium für Forschung und Technologie (BMFT, Publisher) Deutscher Delphi-Bericht zur Entwicklung von Wissenschaft und Technik, Bonn (1993)

Cuhls, K.: From Forecasting to Foresight processes – New participative Foresight Activities in Germany. In: Cuhls, K., Salo, A. (Guest eds.) Journal of Forecasting, Special Issue, no. 22, pp. 93–111. Wiley Interscience (2003)

Cuhls, K., Blind, K., Grupp, H. (Hg.): Delphi 1998 Umfrage. Zukunft nachgefragt. Studie zur globalen Entwicklung von Wissenschaft und Technik, Karlsruhe (1998)

Cuhls, K., Kuwahara, T.: Outlook for Japanese and German Future Technology, Comparing Technology Forecast Surveys. Physica, Heidelberg (1994)

Häder, M., Häder, S.: Die Delphi-Technik in den Sozialwissenschaften. Methodische Forschungen und innovative Anwendungen. Westdeutscher Verlag, Opladen (2000)

Koschatzky, K.: Räumliche Aspekte im Innovationsprozess. Ein Beitrag zur neuen Wirtschaftsgeographie aus Sicht der regionalen Innovationsforschung. In: Wirtschaftsgeographie, vol. 19, Münster, Hamburg (2001)

Leydesdorff, L.: The Triple Helix Model And The Study of Knowledge-based Innovation Systems. International Journal of Contemporary Sociology, Jg. 42, Nr.1 (2005)

Martin, B.R.: Foresight in Science and Technology. Technology Analysis & Strategic Management, Jg.7, Nr.2., S.139–168 (1995)

A Modern Approach to Total Wellbeing

Maja Hadzic[1], Meifania Chen[1], Rick Brouwer[2], and Tharam Dillon[1]

[1] DEBII, Research Lab for Digital Health Ecosystems,
Curtin University of Technology, GPO Box U1987, Perth 6845, WA, Australia
{m.hadzic,m.chen,t.dillon}@curtin.edu.au
[2] Total Wellbeing Medical and Counseling Centre, Suite 1 / 857 Doncaster Rd,
Doncaster East 3109, Victoria, Australia
rickb@totalwellbeing.com.au

Abstract. The events of the last decades have impacted our lives and our health significantly. We expected that the technology boom will improve our lives. While this may be true in a specific context, generally speaking our societies are suffering from moral decays, terrorism fears, wars, financial crisis and unpredictable acts of nature that are increasing in frequency and in intensity. The complex nature of the world we live is impacting our health and wellbeing considerably. Our health is not only determined by our physical health but is the end product of the interplay of the physical, mental, emotional, financial, relational and spiritual events of a lifetime. In this paper we develop a framework that will help us define and measure total wellbeing of individuals in our volatile societies. This framework will help us better understand the complex nature of total wellbeing and develop effective prevention and intervention strategies.

Keywords: total wellbeing, health informatics, ontology, electronic total wellbeing record, data mining.

1 Introduction

We live in a digital milieu, we have a high standard of living but the increase in affluence and materialism has failed to bring us better health, greater inner peace and a fuller sense of meaning, direction and satisfaction [1]. The revolutionary technology development has resulted in rapid introduction of cutting-edge technologies into our societies. We became very dependant on the high technologies and comfort they brought. We started enjoying the modern way of living and were hoping that the new things will make our lives better. We were up for better lives.

While the lives of individuals may have become better, evidence [1] suggests that general health and wellbeing of our societies became worse. Since 1960:

- the divorce rate has doubled
- the teen suicide rate has tripled
- the recorded violent crime rate has quadrupled
- the prison population has quintupled

M. Ulieru, P. Palensky, and R. Doursat (Eds.): IT Revolutions 2008, LNICST 11, pp. 140–150, 2009.
©ICST Institute for Computer Sciences, Social-Informatics and Telecommunications Engineering 2009

- the percentage of the babies born to unmarried parents has increased six fold
- cohabitation (a predictor of future divorce [2]) has increased sevenfold

Terrorist attacks and fear of terrorism are adding additional pressure on our societies [3]. Wars and rumors of wars are marking our society. This moral corruption is closely followed by the financial insecurity, times of market instabilities, increasing interest rates and bankruptcies [4].

We have control of the abovementioned factors but have still allowed them to degrade our societies, lives and health. We are exposed to additional pressure created by the factors we have less control of. For example, earthquakes, tsunamis, tornados and cyclones have not only increased in frequency but also in intensity. There were 368 documented US tornados in January and February this year, which exceeded the previous record of 243 in 1999 for that two month period [5]. The recent cyclone in Burma and earthquakes in China have shaken both nations.

It appears that the occurring problems are increasing over time, and are gaining a momentum rather than being random events. The increasing stress, pressure and fears associated with these events are affecting our lives negatively. Depression, as predicted, will be the world's leading cause of disability by 2020 [6]. Why do we live more modernized but not necessarily happier lives as we enter the twenty-first century?

In this paper we focus on developing a framework that will help us precisely define and measure the total wellbeing of individuals in our modern society. Total wellbeing is not only determined by our physical health but the end product of the interplay of the physical, mental, emotional, financial, relational and spiritual events of a lifetime. The background of this statement is explained in Section 2. The six-dimensional conceptual model or 6D Ontology is described in Section 3. In Section 4 we illustrate how the 6D Ontology can be used in the design of intelligent health information systems. This paper is concluded in Section 5.

2 Why the Six-Dimensional Approach?

The concept of health as a "state of complete *physical*, *mental* and *social* well-being and not merely the absence of disease or infirmity" was originally defined by the World Health Organization [7]. Since then, a multitude of research has attempted to delineate the factors that contribute to positive human health. In this paper, we will explain that additional dimensions need to be taken into account when defining and accessing total wellbeing of individuals. These include *emotional*, *financial* and *spiritual* dimensions.

A number of researchers have established links between 2 or 3 different aspects of health. The complex relationships between the different aspects of health have been fueled by energetic and innovative research programs in several fields, including sociology, psychology, health behavior and health education, psychiatry, gerontology, and social epidemiology [8].

2.1 Physical and Mental Health

Numerous studies have demonstrated a strong association between mental perceptions and physical health, in particular behavioral responses to stress. Chronic activation of the stress response can put a strain on various organs, leading to systems breakdown, compromised immune response and ultimately the deterioration of physical health [9, 10]. Significant links between mental illnesses such as depression, and chronic physical illnesses such as asthma [11], diabetes [12] and cardiovascular disease [13] have also been reported. According to the 1996 U.S. Surgeon General's Report on Physical Activity and Health [14], physically active people tend to have better mental health. In comparison to inactive people, the physically active had higher scores for positive self-concept, more self-esteem and more positive "moods" and "affects." Consequently, physical activity has been successfully used as a non-pharmacological treatment for depression [15, 16]. Additionally, physical activity may reduce the symptoms of anxiety, improve self-image, social skills, and cognitive functioning, and be a beneficial adjunct for alcoholism and substance abuse programs [17].

2.2 Financial, Physical and Mental Health

Cross-national comparison studies report an elevated level of wellbeing from countries high in per capita income (GNP); in contrast, countries lowest in reported wellbeing were those in Eastern Europe, where people were suffering from low income [18]. A comparison between self-reported happiness and GNP from seven parts of the world indicated a clear relationship between happiness and per capita GNP [19]. However, there are some inconsistencies between the data; Latin America scored almost as high in reported happiness as Western Europe, which has more than double the Latin American GNP. This suggests that even though financial factors play a large role in total wellbeing, there are other factors which play a significant part in contributing to total wellbeing. Data from the National Health Interview Survey, the National Survey of Families and Households, the Survey of Income and Program Participation indicates that increases in income significantly improve mental and physical health [20]. However, the study also provides evidence that increases in income increase the prevalence of alcohol consumption which in its turn may damage mental and physical health.

2.3 Social, Physical and Mental Health

A study by House et al. [21] has provided both a theoretical basis and strong empirical evidence for impact of social relationships on health. Prospective studies exposed the evidence that persons with a low quantity, and sometimes low quality, of social relationships have increased risk of death. Experimental and quasi-experimental studies of humans and animals gave similar results. The social isolation has proven to be the major risk factor for mortality from widely varying causes. While House et al. [21] report increased mortality rate in isolated individuals and Kawachi & Berkman [22] highlight a beneficial role of the social ties play on the mental wellbeing, not all relationships are beneficial. Some relationship can be destructive and it is better to avoid them. Rolland [23] does study from a different perspective. He studies situations where illness or disability strikes a couple's relationship. These situations often put the relationship out of balance and consequently result in dysfunctional relationship patterns.

2.4 Emotions, Physical and Mental Health

The relationship between emotional and physical health has also been explored. It was demonstrated that emotional stress such as anxiety has a negative impact on immunity [24]. In a recent study, anxiety characteristic was even found to be a significant and independent predictor of cardiovascular disease [25]. Dr Colbert [26] further defines the destructive emotions, their origin and manifestations and exposes their negative effect on health. The toxic effect of these emotions may result in a variety of illnesses including hypertension, arthritis, multiple sclerosis, irritable bowel syndrome and some types of cancer.

2.5 Physical and Spiritual Health

Positive association between *physical* health and *spirituality* has been reported by Powell *et al.* [27]. The researchers state that the risk of mortality is reduced by 25% in church/service attendees and conclude that church/service attendance protects healthy people against death. Additionally, they provide evidence that the religion or spirituality protects against cardiovascular disease, which is largely mediated by the healthy lifestyle the doctrine encourages.

2.6 Mental and Spiritual Health

Dr D'Souza [28, 29] highlights the need of *mental health* patients to have their *spiritual* issues addressed. The majority of patients rated spirituality as very important and wanted their therapist to take their spiritual needs into consideration. Sixty-seven per cent of the patients said that their spirituality helped them cope with their condition. As the majority of the mentally ill patients appeared to be spiritual, the question here is: "Did spirituality trigger mental disorder in the first place?" One would expect relationships with the divine to affect the mental health positively. Is it possible that the patients had revelation of God in the past, but have lost the closeness to God over time? Their minds became battlefields of knowing what is right and doing what is wrong which brought continual torment upon their minds, eventually resulting in mental disorder. Bergin [30] addresses these questions as he discusses the effect of spirituality on mental health. He provides some explanation on the multi-factorial nature of religion and describes both positive and negative effects of religion. The positive impact is marked by inspirational conviction or commitment and is manifested in dramatic personal healing or transformation. The effects have proven to be extraordinary when this kind of experience becomes linked with social forces. The negative impact is described as evil clothed in religious language or religion that is clearly not constructive. He states that 'spiritual phenomena have an equal potential for destructiveness, as in fundamentalist hate groups'.

Since the identified factors are interactive and mutually dependant on one another, there is a need to employ an integrated, multidimensional approach to total wellbeing. This requires the analysis of each factor individually and in relation to other factors. The study of a single dimension on its own is not sufficient. We can find no evidence that such comprehensive research has ever been performed i.e., the design of an intelligent system to map the different factors on a person's total wellbeing and reveal the interrelationships and interdependencies between the six dimensions.

We need to fully embrace information technology and its potential to develop an information infrastructure that will enable us to simultaneously study the different dimensions and relationships between them. In this paper, we propose a six-dimensional model that takes into account physical condition, along with mental, emotional, financial, relational and spiritual wellbeing. The Physical, Mental, Emotional, Financial, Relational and Spiritual Wellbeing of a person are the six 'dimensions' (6D) that interact to culminate in total wellbeing.

3 6D Ontology

Ontology is an enriched conceptual model for representing domain knowledge. Ontology captures and represents specific domain knowledge through specification of meaning of concepts including definition of the concepts and domain-specific relationships between those concepts. Ontology provides a shared common understanding of a domain and has been suggested as a mechanism to provide applications with domain knowledge and support knowledge integration, use and sharing by different applications, software systems and human resources [31]. Hence, ontology can be used to empower a system with the required knowledge and enable it to operate effectively and efficiently.

The importance of ontologies has been recognised within the medical community and work has begun on developing and sharing biomedical ontologies [32]. A number of biomedical ontologies exist, for example, Gene Ontology and UMLS. The aim of Gene Ontology (GO) (http://www.geneontology.org/) project is to enable consistent descriptions of gene products in different databases by using GO to annotate major repositories for plant, animal and microbial genomes. The Unified Medical Language System (UMLS) [33] is a collection of many biomedical vocabularies. There are 1 million biomedical concepts in UMLS, as well as 135 semantic types and 54 relationships used to classify these concepts. Human Disease Ontology [34] captures and represents knowledge about human diseases. It consists of disease types, symptoms, causes and treatments subontologies. Protein Ontology (http://proteinontology.info/) [35] provides a unified vocabulary for capturing declarative knowledge about protein domain and to classify that knowledge to allow reasoning. It acts as a mediator for accessing not only related data but also semi-structured data such as XML or metadata annotations and unstructured information. A great variety of biomedical ontologies is available via The Open Biomedical Ontologies (http://obofoundry.org/) covering various domains such as anatomy, biological processes, biochemistry, health and taxonomy.

We are in the process of creating a 6D Ontology (6DO), a comprehensive conceptual model to capture and represent the knowledge specific to the six dimensions of the Total Wellbeing. In Figure 1, we represent Total Wellbeing as a result of Physical, Mental, Emotional, Financial, Relational and Spiritual Wellbeing. 6DO will specify the meaning of the concepts used within total wellbeing domain, including definition of these concepts and the domain-specific relationships between them. These precise specifications will constrain the potential interpretations of 6DO concepts and enable 6DO to be used by automatic application. For the modelling of 6DO, we have utilized Protégé tool developed by Stanford University.

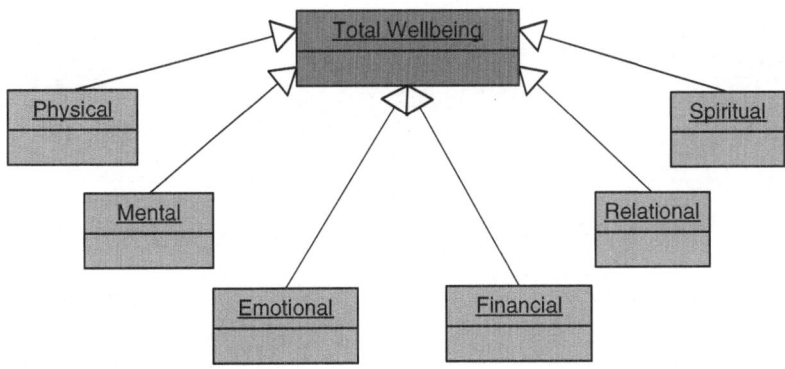

Fig. 1. The Six Dimensions of the Total Wellbeing Ontology

6DO provides a common and shared conceptual framework for specification of each of the Total Wellbeing dimensions. 6DO contains generic information that is always true for the 6D system. Each dimension is further specified by a number of factors that jointly define this particular dimension. For example, the Relational dimension is defined by different kinds of relationships such as with parents, with spouse, with children, with friends, with class mates, with work colleagues, with appointed authorities etc. The emotional wellbeing is determined by different kinds of emotions such as anger, anxiety, bitterness, delusion, disbelief, doubt, fear, guilt, happiness, hope hurt, joy, peace, rage, self-hatred, self-rejection, stress, torment, trauma, etc.

We are told that every person has a unique background, understanding, needs, desires and goals, but are there certain patterns that emerge as we observe the different circumstances in which people live? Doctors have discerned that some emotions are common to patients experiencing same illness. Until now, this has only been an observation. Making a formal model like 6DO will enable us to effectively capture interesting information, and through further data analysis, expose the evidence that will help us manage our health better.

The use of 6DO within health information systems can have many advantages such as (1) increasing data semantic (i.e. provide context for information); (2) enabling of effective knowledge storage, structuring, organization, representation, management, sharing and creation; and (3) supporting intelligent information integration and analysis. In the next section, we will describe a health information system that can be designed to incorporate 6DO.

4 6DO-Based Health Information System

The framework specified by 6DO can be used to design software to systematically collect, store and analyse patients' information. The four different phases are shown in Figure 2.

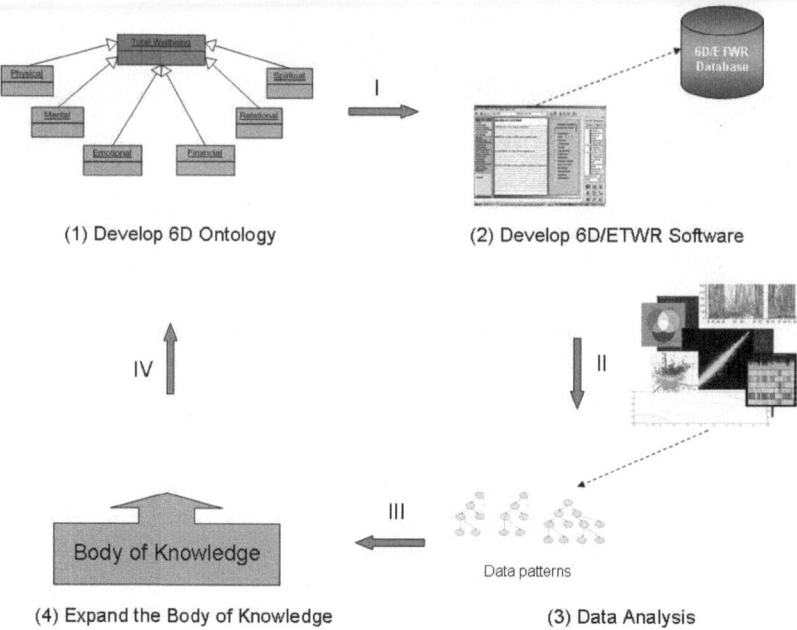

(1) Develop 6D Ontology   (2) Develop 6D/ETWR Software

(4) Expand the Body of Knowledge   (3) Data Analysis

Fig. 2. 6D intelligent information system

In the first phase, the existing body of knowledge is used to design 6DO. We have already started designing the ontology using the Protégé tool. An extensive literature survey is being carried out to identify the key ontology concepts. Additionally, a number of domain experts is involved to advise the ontology design.

In the second phase, 6DO will be used as a basis for design a software that will capture and store patients' information. The 6D data template will be populated by information specific to an individual and saved as Electronic Total Wellbeing Record (ETWR) for each individual. The principle of population of generic ontology by specific information to create specific ontologies is explained in our previous work [36]. The ETWR will contain information describing a person's Physical, Mental, Emotional, Financial, Relational and Spiritual Wellbeing. Privacy preserving mechanisms must be implemented in such system. Our research centre specializes in data privacy issues, particularly privacy of medical data [37, 38].

In the third phase, the collected data will be analysed using automatic analysis techniques such as data mining. Data mining techniques enable us to explore and analyse the collected data, and identify embedded patterns and knowledge in this data. Data mining techniques permit advanced and effective study of complex problems.Within the health domain, data mining techniques have been predominately used for tasks such as analysis of genes and proteins, text mining and drug design. We will apply the data mining to collectively mine ETWRs for noteworthy data patterns. Our research centre has developed a number of data mining algorithms for both structured [39, 40, 41, 42, 43] and semi structured data [44, 45, 46, 47, 48]. The developed algorithms were applied on large and complex data and these experiments successfully demonstrated the scalability of the developed algorithms [39, 41, 43]. From an application perspective in [49], we have applied our tree mining algorithm to extract useful pattern structures from the Protein Ontology database for Human Prion proteins [50].

The 6D dataset under examination will be split into two subsets, one for deriving the knowledge model (source set) and one for testing the derived knowledge model (test set). The source set will be intelligently explored and analysed to extract information, find hidden patterns and knowledge embedded in the data. Predictive models will be proposed. These hypotheses if supported through further testing, could make a significant contribution to research into human Total Wellbeing. The test set will be used to verify the hypothesis so that it can become reliable enough to extend the current knowledge. The research team will collaboratively work and experiment with variation of data mining parameters as their choice can affect the nature and granularity of the obtained results.

In the fourth phase, the knowledge that has been revealed and validated during the data mining phase will be published. This will extend the existing body of knowledge and in the combination with other new knowledge that emerged over time, may be used to extend and/or correct the ontology model. This principle of ontology evolution is frequent in the societies with rapidly accumulating knowledge.

5 Conclusion

Our society is changing in all aspects. The technology breakthrough, the modern way of living and modern societies have not only resulted in comfortable lives but also in increased stress, pressure and fears which in their turn started degrading our health and wellbeing.

More than ever, we need to take a holistic approach in studying and controlling our health ad wellbeing. We have exposed evidence to support the statement that the total wellbeing is the end product of the interplay of the following six dimensions or events of a lifetime: physical, mental, emotional, financial, relational and spiritual. We are currently designing 6DO, an integrated conceptual model to capture and represent each dimensions on its own an in relationship with other dimensions. We aim to design a software based on 6DO that will systematically collect and analyze data to expose the evidence that will support general public in managing their personal wellbeing better, and health professionals in adapting their services to address patients' needs more systematically and effectively. Innovative breakthroughs from this technique will change the way we understand, control and manage our health. A pilot study will be carried out at the Total Wellbeing Medical and Counseling Centre, but generic findings will be made available to the general public, and health providers both in policy and practice that target total wellbeing.

References

1. Myers, D.G.: The American paradox: Spiritual hunger in an age of plenty. Yale University Press (2000)
2. Hall, D.R., Zhao, J.Z.: Cohabitation and Divorce in Canada: Testing the Selectivity Hypothesis. Journal of Marriage and the Family 57, 421–427 (1995)

3. Hoffman, B.: Al Qaeda, Trends in Terrorism, and Future Potentialities: An Assessment. Studies in Conflict and Terrorism 26(6), 429–442 (2003)
4. Fedor K. AAP: Financial crisis: Fears of world slump after Lehman Brothers crash, http://livenews.com.au/Articles/2008/09/16/Financial_crisis_Fears_of_world_slump_after_Lehman_Brothers_crash_
5. Mooney, C.: Our Extraordinary Tornado Year, http://www.thedailygreen.com/environmental-news/blogs/hurricanes-storms/tornado-trends-55030501
6. Lopez, A.D., Murray, C.C.J.L.: The Global Burden of Disease, 1990-2020. Nature Medicine 4, 1241–1243 (1998)
7. World Health Organization: Preamble to the constitution of the world health organization as adopted by the international health conference, New York (June 19–22 , 1946); signed on July 22, 1946 by the representatives of 61 States (Official Records of the World Health Organization, no. 2, p. 100) and entered into force on April 7, 1948
8. Ellison, C.G., Boardman, J.D., Williams, D.R., Jackson, J.S.: Religious involvement, stress, and mental health: Findings from the 1995 Detroit area study. Social Forces 80(1), 215–249 (2001)
9. Weiss, J.M.: Stress-induced depression: Critical neurochemical and electrophysiological changes. In: Madden, J. (ed.) Neurobiology of learning, emotion, and affect, pp. 123–154. Raven Press, New York (1991)
10. Sapolsky, R.M.: Stress, glucocorticoids, and damage to the nervous system: The current state of confusion. Stress 1, 1–19 (1996)
11. Strine, T.W., Mokdad, A.H., Balluz, L.S., Berry, J.T., Gonzalez, O.: Impact of depression and anxiety on quality of life, health behaviors, and asthma control among adults in the United States with asthma. Journal of Asthma 45(2), 123–133 (2008)
12. Strine, T.W., Mokdad, A.H., Dube, S.R., Balluz, L.S., Gonzalez, O., Berry, J.T., Manderscheid, R., Kroenke, K.: The association of depression and anxiety with obesity and unhealthy behaviors among community-dwelling US adults. General Hospital Psychiatry 30(2), 127–137 (2008)
13. Rozanski, A., Blumenthal, J.A.,, Kaplan, J.: Impact of psychological factors on the pathogenesis of cardiovascular disease and implications for therapy. Circulation 99, 2192–2217 (1999)
14. U.S. Surgeon General's Report on Physical Activity and Health (1996)
15. Phillips, W.T., Kiernan, M., King, A.C.: Physical Activity as a Nonpharmacological Treatment for Depression: A Review. Complementary Health Practice Review 8(2), 139–152 (2003)
16. Pilu, A., Sorba, M., Hardoy, M.C., Floris, A.L., Mannu, F., Seruis, M.L., Velluti, C., Carpiniello, B., Salvi, M., Carta, M.G.: Efficacy of physical activity in the adjunctive treatment of major depressive disorders: preliminary results. Clinical Practice and Epidemiology in Mental Health 3(8) (2007)
17. Taylor, C.B., Sallis, J.F., Needle, R.: The relation of physical activity and exercise to mental health. Public Health Report 100(2), 195–202 (1985)
18. Inglehart, R.: Culture shift in advanced industrial society. Princeton University Press, Princeton (1997)
19. Veenhoven, R.: Questions on happiness: Classical topics, modern answers, blind spots. In: Strack, F., Argyle, M., Schwarz, N. (eds.) Subjective well-being: An interdisciplinary perspective, pp. 7–26. Pergamon Press, Oxford (1991)
20. Ettner, S.: New evidence on the relationship between income and health. Journal of Health Economics 15(1), 67–85 (1996)
21. House, J.S., Landis, K.R., Umberson, D.: Social relationships and health. Science 241(4865), 540–545 (1988)
22. Kawachi, I., Berkman, L.F.: Social ties and mental health. Journal of Urban Health 78(3), 458–467 (2001)

23. Rolland, J.S.: In Sickness and in Health: the Impact of Illness on Couples Relationships. Journal of Marital and Family Therapy 20(4), 327–347 (2007)
24. Fleshner, M., Brohm, M., Watkins, L.R., Laudenslager, M.L., Maier, S.F.: Modulation of the in vivo antibody response by. benzodiazepine-inverse agonist (DMCM) given centrally or peripherally. Physiol. Behav. 54, 1149–1154 (1993)
25. Shen, B.J., Avivi, Y.E., Todaro, J.F., Spiro, A., Laurenceau, J.P., Ward, K.D., Niaura, R.: Anxiety characteristics independently and prospectively predict myocardial infarction in men the unique contribution of anxiety among psychologic factors. Journal of the American College of Cardiology 51(2), 113–119 (2008)
26. Colbert, D.: Deadly emotions: understand the mind-body-spirit connection that can heal or destroy you. Thomas Nelson (2003)
27. Powell, L.H., Shahabi, L., Thoresen, C.E.: Religion and Spirituality: Linkages to Physical Health. American Psychologist 58(1), 36–52 (2003)
28. D'Souza, R.: Do patients expect psychiatrists to be interested in spiritual issues? Australasian Psychiatry 10(1), 44–47 (2002)
29. D'Souza, R.: Incorporating a spiritual history into a psychiatric assessment. Australasian Psychiatry 11(1), 12–15 (2003)
30. Bergin, A.E.: Values and Religious Issues in Psychotherapy and Mental Health. American Psychologist 46(4), 394–403 (1991)
31. Gómez-Pérez, A.: Knowledge sharing and reuse. In: The Handbook on Applied Expert Systems, pp. 1–36. CRC Press, Boca Raton (1998)
32. Ceusters, W., Martens, P., Dhaen, C., Terzic, B.: LinkFactory: an advanced formal ontology management System. In: Interactive Tools for Knowledge Capture (KCAP 2001) (2001)
33. Bodenreider, O.: The Unified Medical Language System (UMLS): integrating Biomedical terminology. Nucleic Acids Research 32(1), 267–270 (2004)
34. Hadzic, M., Chang, E.: Ontology-based Support for Human Disease Study. In: Hawaii International Conference on System Sciences (HICSS38 2005) (2005)
35. Sidhu, A.S., Dillon, T.S., Chang, E.: Integration of Protein Data Sources through PO. In: Bressan, S., Küng, J., Wagner, R. (eds.) DEXA 2006. LNCS, vol. 4080, pp. 519–527. Springer, Heidelberg (2006)
36. Hadzic, M., Chang, E.: Ontology-based Multi-agent Systems Support Human Disease Study and Control. In: Czap, H., Unland, R., Branki, C., Tianfield, H. (eds.) An Introduction to the PL/CV2 Programming Logic, vol. 135, pp. 129–141. IOS Press, Amsterdam (2005)
37. Hecker, M., Dillon, T.S.: Ontological Privacy Support for the Medical Domain. In: National e-Health Privacy and Security and Symposium, Australia (2006)
38. Hecker, M., Dillon, T.S.: Privacy support and evaluation on an ontological basis. In: IEEE 23rd International Conference on Data Engineering, Turkey (2007)
39. Hadzic, F., Dillon, T.S.: CSOM: Self Organizing Map for Continuous Data. In: International Conference on Industrial Informatics, Australia (2005)
40. Hadzic, F., Dillon, T.S.: Using the Symmetrical Tau (τ) Criterion for Feature Selection in Decision Tree and Neural Network Learning. In: International Workshop on Feature Selection for Data Mining: Interfacing Machine Learning and Statistics, in conj. with SIAM 2006, USA (2006)
41. Hadzic, F., Dillon, T.S.: CSOM for Mixed Data Types. In: Liu, D., Fei, S., Hou, Z., Zhang, H., Sun, C. (eds.) ISNN 2007. LNCS, vol. 4492, pp. 965–978. Springer, Heidelberg (2007)
42. Hadzic, F., Dillon, T.S., Tan, H.: Outlier detection strategy using the Self-Organizing Map. Knowledge Discovery and Data Mining: Challenges and Realities with Real World Data (2006)
43. Hadzic, F., Dillon, T.S., Tan, H., Feng, L., Chang, E.: Mining Frequent Patterns using Self-Organizing Map. Advances in Data Warehousing and Mining Series (2006)

44. Hadzic, F., Tan, H., Dillon, T.S.: UNI3 – Efficient Algorithm for Mining Unordered Induced Subtrees Using TMG Candidate Generation. In: IEEE Symposium on Computational Intelligence and Data Mining, USA (2007)
45. Tan, H., Dillon, T.S., Hadzic, F., Chang, E., Feng, L.: MB3-Miner: mining eMBedded sub-TREEs using Tree Model Guided candidate generation. In: International Workshop on Mining Complex Data, held in conjunction with ICDM 2005, USA (2005)
46. Tan, H., Hadzic, F., Dillon, T.S., Feng, L., Chang, E.: Tree Model Guided Candidate Generation for Mining Frequent Subtrees from XML. ACM Transactions on Knowledge Discovery from Data (2008)
47. Tan, H., Dillon, T.S., Hadzic, F., Chang, E.: Razor: mining distance constrained embedded subtrees. In: IEEE ICDM 2006 Workshop on Ontology Mining and Knowledge Discovery from Semistructured documents, China (2006)
48. Tan, H., Dillon, T.S., Hadzic, F., Feng, L., Chang, E.: IMB3-Miner: Mining Induced/Embedded subtrees by constraining the level of embedding. In: Pacific-Asia Conference on Knowledge Discovery and Data Mining (2006)
49. Hadzic, F., Dillon, T.S., Sidhu, A.S., Chang, E., Tan, H.: Mining Substructures in Protein Data. In: IEEE ICDM DMB Workshop, China (2006)
50. Sidhu, A.S., Dillon, T.S., Sidhu, B.S., Setiawan, H.: A Unified Representation of Protein Structure Databases. In: Reddy, M.S., Khanna, S. (eds.) Biotechnological Approaches for Sustainable Development, pp. 396–408 (2004)

Applying Business Process Re-engineering Patterns to optimize WS-BPEL Workflows

Jonas Buys, Vincenzo De Florio, and Chris Blondia

University of Antwerp, PATS research group,
Middelheimlaan 1, B-2020 Antwerp, Belgium
{jonas.buys,vincenzo.deflorio,chris.blondia}@ua.ac.be

Abstract. With the advent of XML-based SOA, WS-BPEL shortly turned out to become a widely accepted standard for modeling business processes. Though SOA is said to embrace the principle of business agility, BPEL process definitions are still manually crafted into their final executable version. While SOA has proven to be a giant leap forward in building flexible IT systems, this static BPEL workflow model is somewhat paradoxical to the need for real business agility and should be enhanced to better sustain continual process evolution. In this paper, we point out the potential of adding business intelligence with respect to business process re-engineering patterns to the system to allow for automatic business process optimization. Furthermore, we point out that BPR macro-rules could be implemented leveraging micro-techniques from computer science. We present some practical examples that illustrate the benefit of such adaptive process models and our preliminary findings.

Keywords: business process re-engineering (BPR), service-oriented architectures (SOA), WS-BPEL, business processes, workflows.

1 Introduction

A cutthroat competition between enterprises is currently raging in which companies are compelled to constantly evolve so as to realize a competitive advantage. This head start is pursued by iteratively altering business processes[1] and strategies aimed at improving operational efficiency [1]. Business processes are thus continuously refined, mainly to resolve recurrent issues and as such rectify process performance. This concept is commonly referred to as business process re-engineering (BPR)[2].

---

[1] The notion of business process is defined as an orchestration of several activities carried out by computer systems or people within an enterprise with the objective of supplying a product or service to the customer.

[2] Because of the vague definitions found in most text books, the BPR acronym is commonly used interchangeably for business process re-engineering as well as business process redesign. The former has an evolutionary character, while the latter is revolutionary. For more information, we refer to [1].

M. Ulieru, P. Palensky, and R. Doursat (Eds.): IT Revolutions 2008, LNICST 11, pp. 151–160, 2009.

Information technology (IT) is actively deployed in large enterprises and this also applies to process automation. But regrettably most of these volatile business processes are enlaced in rigid IT systems and this imposes limitations with respect to the speed changes are possible with. In the beginning of this decade, this issue led to the concept of service-oriented architectures (SOA) in which IT is flexibly structured to better alleviate the re-engineering of processes by splitting up so-called business logic in a number of software components that are exposed as services [3]. Service (operations) should correspond to individual process activities that processes can be composed of by orchestrating and coordinating functionality comprised in those services. Actually this service-oriented computing paradigm comprises the best practices in distributed computing of - roughly estimated - the past twenty years, and commercially backed by major industry concerns, SOA keeps gaining popularity [4].

As one possible SOA implementation technology, web services have managed to become the *de facto* standard for enterprise software in which various distributed, heterogeneous software systems are integrated in support of corporate e-business and e-commerce activities [3]. The Web Services Business Process Execution Language (WS-BPEL) XML language is one of the standards that resulted from intensive standardization initiatives of industrial consortia, and shortly became a widely accepted standard for workflow modeling [6]. The benefit of the central BPEL orchestration component is that the process definition is no longer interwoven in the implementation code of the business logic. Because of this separation, SOA is said to alleviate the transformation and restructuring of business processes using highly reusable services that can easily be re-orchestrated into BPEL workflows [3].

The service-oriented paradigm turned out to be a giant leap forward in the construction of flexible IT systems indeed. XML-based SOA with BPEL further added to business agility, allowing for the quick development of new business processes leveraging service-wrapped legacy IT assets (*i.e.* business process redesign). But in spite of the popularity of BPEL and its clear separation of process and business logic, some shortcomings are not resolved thus far [5]. One of these issues is that a BPEL process definition is uttermost static: it is designed manually using some software tools and is then loaded into the BPEL engine. Since service orchestration and business processes are at the centre of SOA, it is imperative to continuously optimize BPEL process definitions to achieve system performance, besides the economic point of view in realizing a competitive advantage. Hence, this static BPEL workflow model should be enhanced to better sustain business process agility. Failing to do so would result in a somewhat paradoxical situation in which this static BPEL model is expected to drive the inherently dynamic processes required by the actual continual process evolution.

Although the BPR methodology originated in the early Nineties, until recently, businesses were still generally managed using an approach based on experience and intuition. But as BPR is gaining adherence, we are on the verge of uniting the IT-driven service-oriented paradigm and the BPR managerial methodology: automatically applying prevailing BPR principles to BPEL process definitions can help in the further optimization of these process models, therefore sustaining process evolution.

One possibility to combining both disciplines is to build BPR intelligence into workflow design tools, transforming the process model before it is loaded into the BPEL engine. Alternatively, attributing business intelligence to SOA allows for the dynamic application of BPR principles to the original static BPEL process definitions with the goal of optimizing the process at runtime with respect to the system's current state and execution environment, and also allows to computerize more complex BPR patterns.

Leveraging techniques and practices from computer science to implement BPR patterns, adaptive process models arise, so that a more advanced form of SOA business agility can be attained. The ability to dynamically modify business processes such that the process semantics are preserved, while at the same time the process model is optimized, will positively impact SOA performance and add to the operational efficiency of a company striving for a competitive lead.

In this paper we illustrate the applicability of BPR patterns to BPEL workflow processes, and show how this can lead to performance improvements, such as a reduction in execution time.

2 Business Process Re-engineering and BPEL

Numerous best practices (principles, design patterns, heuristics) for BPR have been proposed in the literature [2], yet there has not been any thorough inquiry into combining IT and BPR so far. In order to support a higher level of process agility, we propose to design an intelligent system able to optimize the BPEL processes in accordance with these conceptual BPR principles. An overview of some potentially useful patterns for BPEL process improvement is shown in table 1.

Table 1. Some business process re-engineering patterns

BPR directives	Basic techniques	Human-centric
resequencing	data and control flow analysis [7],	No
parallelization	Tomasulo, scoreboarding [8]	No
exception	control flow and flow variable	No
knock-out (minimize process cost)	speculation [8]	No
order assignment and distribution	*chain of execution* [10]	Yes
flexible assignment	*nomination* [10,11]	Yes
specialist-generalist		Yes
split responsibilities	*4-eyes principle* [10]	Yes

The WS-BPEL XML language defines a set of primitives which business processes can be modeled with: basic activities can be set in order using control and data flow supported by structured activities [6,3]. These rudimentary structural activities turn out to be limited to the common control and decision structures available in most imperative programming languages. It is no surprise,

then, that we can spot similarities between the techniques for program optimization in computer science, operating on atomic units of instructions (which we
refer to as the micro-level domain, or *micro* for short) and the BPR patterns for
business processes, acting on coarser units of instruction blocks with a variable
size: service operations (*macro*). Consequently, it is plausible to try and automate
economic BPR patterns leveraging existing techniques from computer science.

Computerizing BPR turns out to be twodimensional: on the one hand, offline
optimization techniques can preprocess the process model prior to its execution;
on the other hand, runtime information can be used to adjust the BPEL process
online depending on the system's current state and resource availability, either
by affecting the process model or individual process instances[3].

In the following subsections, some examples will be shown to illustrate the
benefit of applying BPR patterns to BPEL processes, and some suggestions
for possible computer science techniques or practices for the implementation of
these patterns will be motivated. For the purpose of clarity, the examples are
presented in business process modeling notation (BPMN), a common graphical
representation of the actual XML BPEL definition. This does not limit our
contribution, as BPEL can easily be mapped to BPMN and vice versa [12].

2.1   Resequencing and Parallelization BPR Patterns

The execution of a BPEL process is essentially sequential, though the BPEL
specification also provides constructs for executing activities in parallel [6]. As
a primary BPR pattern, the execution order of process activities - *i.e.* service
invocations - can be optimized by considering data flow dependencies so as to
execute mutually independent activities in parallel [2]. The underlying idea of
simultaneously executing activities and advancing activity initialization is that
some time can be gained by avoiding performance-degrading stalls caused by
dependencies. Throughout this paper, we assume an SOA-based environment
aggregating a set of distributed IT systems allowing for optimization by parallel
execution, but the amount of parallelism that can actually be achieved is limited
by the number of resources (web services or employees) in the system.

Data dependencies can arise between different activities and relate to variables
defined in the BPEL process definition. For instance, a service invocation activity has a read-dependency on whatever variable is used to hold the input
message for the service to be invoked, and also a write dependency to the
variable that will ultimately store the service's reply. Assignment statements
normally construct and write to a variable after reading values from one or more
other variables. Structural activities may also read certain variables during the
evaluation of control flow variables.

Computer science techniques for dynamic instruction scheduling such as the
Tomasulo approach and scoreboarding allow for an optimized, out-of-order

---

[3] BPEL processes are usually complex long-running processes. Every time the process
model is triggered, the BPEL engine will construct a new process instance that runs
in isolation from the other instances.

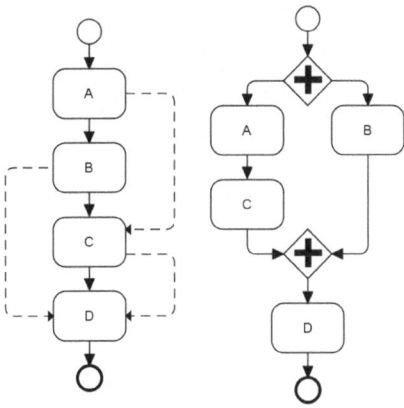

Fig. 1. On the left the original BPEL; on the right the optimized process. The rhombic symbols stand for parallel execution, *i.e.* AND-split/join.

execution of sequential streams of program instructions and could be used as the basis for individual process instances [8]. These techniques could be extended and applied to BPEL activities to avoid pointless waiting as the result of data dependencies. However, these approaches are limited to basic blocks of non-branching sequences of instructions - *cf.* one sequential scope in BPEL - and can benefit from techniques such as speculation to work around control flow statements and as such artificially increasing the number of instructions in these basic blocks. The Tomasulo approach exploits the knowledge on dependencies unravelled at runtime; thus it clearly outperforms all strategies that statically analyze the data and control flow of the BPEL process; nevertheless, such static analysis is easier to build up and can be used to restructure the overall BPEL process model [7].

In figure 1, we merely represent service invocation activities and assume that an unaltered output message from a particular service invocation is stored in a BPEL variable, which is used as input for invoking another service. Note that the dashed arrows representing these data dependencies are not part of the BPMN notation. The start event corresponds to the reception of a message that triggers execution of a BPEL process instance and the end event represents the process replying to its requester. The solid arrows in the diagrams indicate sequential flow. Supposing the respective execution times for activities (services) A, B, C and D are 10, 5, 20 and 7 seconds, the execution time in the original process would simply be 42 seconds, while the optimized version would result in an execution time of 37 seconds.

2.2 Exception and Knock-Out BPR Patterns

In re-engineering business processes, it is common to isolate exceptional flow from the normal process flow. Techniques like speculation are already applied to improve control flow, branches in particular [8]. This can be accomplished by conditionally executing the branch with highest probability, and compensating in case of misprediction. Moreover, the amount of parallelism that one can exploit is also limited by control dependencies. Speculation is a technique that can be used to overcome the penalty of control dependencies in some cases by shifting highly probable activities to eliminate control dependencies so as to match the parallelism offered by the execution environment. To achieve speculation techniques in BPEL process definitions, scopes of activities might be shifted, provided that an estimation on the probability of each branch is available. This information must be gathered at runtime, during the execution of the program, using monitoring approaches as in feedback loops and autonomic computing [9]. An additional gain in execution time can be harvested by applying this pattern to a process because of further parallelization.

Consider for instance the example in figure 2. Imagine the left branch in the original process model has a 40% chance of being executed. For the service invocations A, B, C and X we suppose execution times being 10 seconds each. Then process execution would take 30 seconds for sequentially executing activities X, A and B, or 20 seconds for the other execution path comprising activities X and C. The time required to evaluate control flow structures, *e.g.* branch variables, is considered negligible. Speculation should restructure the model to always execute activity C (because of the 60% chance the containing branch is chosen) in parallel to X. The branching condition remains identical as long as the result of invoking service C is not written to a variable that is used for evaluating the branching condition. In case the execution proceeds accordingly, there is a considerable improvement: 10 seconds instead of 20 seconds in the original model (in this case a speedup of factor 2). However, speculation might deteriorate and delay process execution if the execution does not fit: because of the undoing, the best-case execution time of the speculated BPEL process is now 30 seconds. In the worst case, the performance degradation is given by

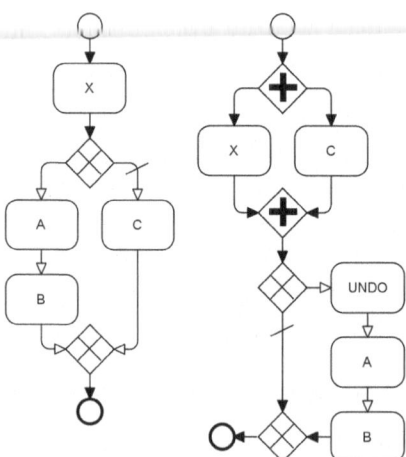

Fig. 2. On the left the original process; on the right the process optimized by speculation. The rhombic symbols in the original process denote conditional branches.

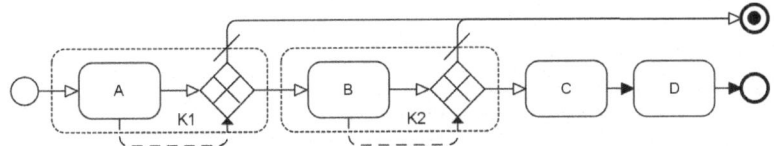

Fig. 3. An example to illustrate the knock-out BPR pattern

the overhead for undoing. Therefore, we suggest speculation to be applied only if the difference in probabilities of the alternative branches exceeds a minimal threshold. In that case, the overall performance might benefit from speculation, and the negative impact of non-fitting BPEL instances might be subdued.

The knock-out BPR pattern, another application where speculation might fit in for implementation, is a special version of the resequencing pattern aiming to manipulate the process yielding on average the least costly execution. Knock-outs, conditional checks that may cause the complete process instance to cease, skipping all subsequent process activities, should be arranged in decreasing order of effort and increasing order of termination probability [2]. Upon occurrence, the BPEL process instance should be abandoned, possibly compensating in order to reverse the service invocations that were required for evaluating the knockout conditions. We illustrate this principle in figure 3. Suppose knock-out condition K_1 has a 40% probability of evaluating negatively and it takes 2 seconds to invoke service A and compute this branching condition. Likewise, for knock-out K_2 these values respectively equal 60% and 4 seconds including the invocation of service B. Then the ratio 0.40/2 for K_1 is higher then 0.60/4 for K_2. Hence, the arrangement of the knock-outs in this process model is fine.

As time goes by and more process instances have been executed, for both re-engineering patterns, the estimated probabilities of the branching conditions might change, which in turn might trigger new changes to the process model at runtime, resulting in adaptive business processes.

2.3   Human-Centric BPR in People-Oriented Business Processes

Human interactions frequently occur in business processes for the manual execution of tasks, *e.g.* an approval [10]. Expenses resulting from employment of people are still a major cost factor in enterprises. Therefore, an efficient allocation of the staff is imperative, and IT can also help to achieve this goal.

This brings us to another shortcoming of BPEL, which is rightly accused of being too automation-centric since it lacks the recognition of employees in process workflows [6,10]. The WS-BPEL4People and WS-HumanTask specification drafts - recently submitted to OASIS for ratification - allow for hybrid SOA in which human actors occur next to customary IT systems exposed as web services [10,11]. Consequently, we propose to automatically apply human-centric BPR patterns to people-centric BPEL4People processes. To our knowledge, this idea has not been previously investigated. Table 1 shows a few of these patterns

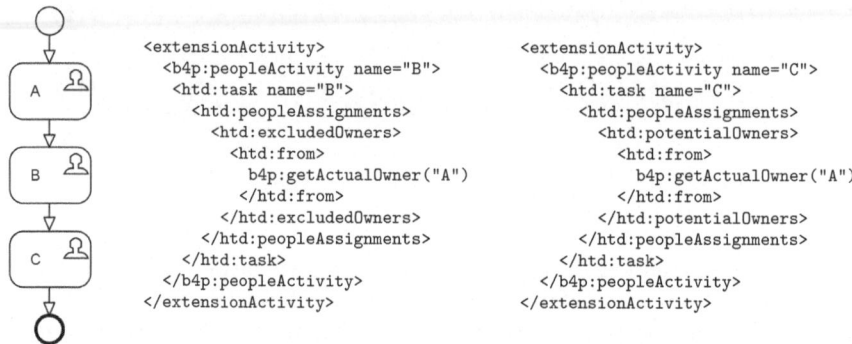

```
<extensionActivity>                        <extensionActivity>
  <b4p:peopleActivity name="B">              <b4p:peopleActivity name="C">
    <htd:task name="B">                        <htd:task name="C">
      <htd:peopleAssignments>                    <htd:peopleAssignments>
        <htd:excludedOwners>                       <htd:potentialOwners>
          <htd:from>                                 <htd:from>
            b4p:getActualOwner("A")                    b4p:getActualOwner("A")
          </htd:from>                                </htd:from>
        </htd:excludedOwners>                      </htd:potentialOwners>
      </htd:peopleAssignments>                   </htd:peopleAssignments>
    </htd:task>                                </htd:task>
  </b4p:peopleActivity>                       </b4p:peopleActivity>
</extensionActivity>                       </extensionActivity>
```

Fig. 4. Sample process comprising human tasks. At the right an excerpt from the BPEL4People process definition illustrating the language constructs for the concepts of separation of duties and chain of execution.

and points out relevant procedures defined in the BPEL4People specifications. Most of them deal with the issue of assigning the task to the best suitable person.

The *order assignment pattern* prefers the same employee to work on several process activities for a particular process instance. This is directly supported by the BPEL4People concept of chain of execution, where the actual owner that took care of the previous activity is selected as the sole potential owner for the task at hand (see figure 4, activities A and C). In addition, an escalation action should be defined to offer the task to the regular set of potential owners in case the default scenario would fail (*i.e.* the owner of the previous activity does not claim the task before the expiration deadline). Assigning several consecutive process activities to one person should result in a reduction of execution time as this person has got acquainted with the case. The side effect, however, is that the employee's workload will slightly increase compared to his or her colleagues.

Next, according to the *flexible assignment BPR pattern*, and supplemented by the *specialist-generalist pattern*, one should distinguish between highly specialized human resources and generalist employees that can be assigned to execute a diversity of tasks. The availability of generalists adds more flexibility to the business process and can lead to a better utilization of resources. Unfortunately, the generic human roles defined in the specifications are insufficient, and the people query facility and the organizational people directory that is searched by this query, both proposed in the above mentioned specifications, remain undefined. We should find a way to annotate people in this directory describing their skills, capabilities and permissions so that the system can reason about the degree of an individual's specialization. We envision an important role for techniques such as semantic processing and semantic matching in particular [13].

Lastly, assigning different tasks within a process to people from different functional units should be avoided (*split responsibilities pattern*). Again, enhancing the expressiveness of people queries and the structure of the people directory could allow the system to optimize the dispatching of human tasks to the appropriate available human resources at runtime. Related to this pattern is the concept of separation of duties, also referred to as the 4-eye principle in which mutually independent individuals each perform an instance of the same task for the purpose of combating fraud and avoiding disastrous mistakes [11]. An example is shown in figure 4, where activity B may not be executed by whoever performed task A.

Combining human-computer interaction by means of the BPEL4People and WS-HumanTask specifications with BPR patterns for automatically and intelligently dispatching workload to human system resources is an exciting research challenge with the potential for a substantial performance improvement in process execution, which may lead to increased productivity and competitiveness. At the same time this also endorses the significance of BPEL4People in SOA, and we plead for a speedy ratification of the specification drafts.

3   Conclusion

We started this paper by briefly introducing BPR as a relatively new managerial methodology and SOA as a way to sustain the volatility of business processes resulting from the fierce competition in the market. It was pointed out that the static nature of BPEL process definitions is somewhat paradoxical with the necessity to quickly and easily re-engineer business processes in the quest towards operational efficiency. With SOA strongly embracing the principle of business agility, incorporating BPR system intelligence into the BPEL engine allows for the dynamic application of BPR principles to the original static BPEL process definitions with the goal of optimizing the process at runtime with respect to the system's current state and runtime environment. We propose an innovatory approach in which BPR principles are explicitly applied to WS-BPEL processes by means of established techniques and practices from computer science so that the process semantics are preserved and whereby at the same time the process execution is being optimized. It is expected that this will allow for a reduction in execution time, *e.g.* as the result of parallelization. This conjecture is corroborated by several small examples. Furthermore, we believe the new WS-BPEL4People standard might allow for the automation of human-centric BPR patterns in people-oriented business processes. This enables the system to intelligently dispatch human tasks to suitable human process actors.

Such BPR-aware SOA have the potential to turn static BPEL process definitions into adaptive workflows matching the current environmental and systemic conditions so as to make a more efficient use of these system resources in view of higher performance. The WS-BPEL specification need not be modified, which ensures a smooth transition in adopting these ideas. Complex BPR patterns can be implemented using runtime system information and possibly service metadata and annotations.

We are still in the early phase of elaborating on the ideas presented in this paper. We intend to develop a proof of concept illustrating the feasibility of the exemplified BPR patterns. Research on the introduced human-centric BPR patterns will depend on the ratification process of the BPEL4People specification draft and the availability of compliant implementations. Moreover, it is worth remarking that the method proposed in this paper does not solely focus on higher performance, but could be extended to focus on other goals such as reliability and availability.

We conclude that BPR-aware SOA environments, automatically applying reengineering patterns to BPEL processes, result in adaptive business processes, which is a crucial requisite to achieve an enhanced form of business agility, better sustaining process evolution and as such mitigating the perceived paradox.

References

1. Reldin, P., Sundling, P.: Explaining SOA service granularity - how IT-strategy shapes services. Master's thesis, Linköpings Universitet, Sweden (2007)
2. Reijers, H.A., Liman Mansar, S.: Best practices in business process redesign - An overview and qualitative evaluation of successful redesign heuristics. Omega 33, 283–306 (2004)
3. Erl, T.: Service-Oriented Architecture: Concepts, Technology, and Design. Prentice Hall, Englewood Cliffs (2005)
4. Stal, M.: Using Architectural Patterns and Blueprints for Service-Oriented Architecture. IEEE Software 23, 54–61 (2006)
5. Buys, J.: Business Process Re-engineering Leveraging Autonomic Service-oriented Computing. Technical report (unpublished) (2008)
6. Abode, et al.: Web Services Business Process Execution Language Version 2.0 (2007)
7. Klopp, O., Khalaf, R.: Reaching Definitions Analysis Respecting Dead Path Elimination Semantics in BPEL Processes. Technical report (2007)
8. Hennessy, J.L., Patterson, D.A.: Computer Architecture - A Quantitative Approach. Morgan Kaufmann Publishers Inc., San Francisco (1990)
9. Ganek, A.G., Corbi, T.A.: The dawning of the autonomic computing era. IBM Systems Journal 42, 5–18 (2003)
10. IBM, SAP, et al.: WS-BPEL Extension for People Specification, Version 1.0 (2007)
11. IBM, SAP, et al.: Web Services Human Task Specification, Version 1.0 (2007)
12. Stephen, A.: Using BPMN to Model a BPEL Process (2005)
13. Sun, H., et al.: Service Matching in Online Community for Mutual Assisted Living. In: Proceedings of the 2007 Third International IEEE Conference on Signal-Image Technologies and Internet-Based Systems (2007)

Applying Semantic Web Services and Wireless Sensor Networks for System Integration

Gian Ricardo Berkenbrock, Celso Massaki Hirata,
Frederico Guilherme Álvares de Oliveira Júnior,
and José Maria Parente de Oliveira

Instituto Tecnológico de Aeronáutica, S. J. Campos, Brazil
{gian,hirata,fred,parente}@ita.br

Abstract. In environments like factories, buildings, and homes automation services tend to often change during their lifetime. Changes are concerned to business rules, process optimization, cost reduction, and so on. It is important to provide a smooth and straightforward way to deal with these changes so that could be handled in a faster and low cost manner. Some prominent solutions use the flexibility of Wireless Sensor Networks and the meaningful description of Semantic Web Services to provide service integration. In this work, we give an overview of current solutions for machinery integration that combine both technologies as well as a discussion about some perspectives and open issues when applying Wireless Sensor Networks and Semantic Web Services for automation services integration.

Keywords: Wireless Sensor Networks, Integration, Semantic Web Services.

1 Introduction

Nowadays changes happen in a faster way. Companies always try to predict when such changes will happen. Buildings, factories, and homes are built to serve for different purposes during their lifetime. And for each purpose we have a distinct requirements set. From this perspective, automation services will change during their lifetime in such environments. Changes are concerned to business rules, process optimization, cost reduction, and so on. And some common changes are re-layout of production line to make a new product or to optimize the process, factory modernization, production lines adjustment for customized products, adaptation to a new layout on a building floor, adjustment for new security and safety rules, owner changes, and so on.

Hence a smooth and straightforward way to deal with machinery automation integration could be using a combining approach of wireless sensor networks and semantic web services. Combining both technologies we could reach an integration among machineries, and between machineries and third-party software systems. The term third-party software is used for software systems that are not deployed in sensor nodes.

M. Ulieru, P. Palensky, and R. Doursat (Eds.): IT Revolutions 2008, LNICST 11, pp. 161–170, 2009.

Wireless sensor networks (WSN) are composed of variety number of small, low cost, and dispensable devices called as sensor nodes or just nodes and these nodes can communicate over-the-air to each other using an embedded radio. The network size can change from a few nodes to hundreds of thousand nodes. WSN have some constraints such as low processing capability, limited energy, low transmission rate, and short transmission range. Due to these constraints WSN are very application-oriented. WSN communication can be single or multi hop. Then to increase communication capability and efficiency many protocols were proposed.

Semantic Web Services (SWS) are services with semantic ontology-based annotations. Contrary to syntactic web services, where parameters are basically described with data types, semantic web services parameters are associated with concepts described by ontologies allowing discovery, composition, execution and monitoring automation of services. By composing existing simple services, it is possible to build dynamic applications in order to perform more complex task to meet requirements. For instance, a complex query (task) that needs to obtain information from several sensors (services).

Therefore, applying together WSN and SWS could give us a smooth way to deliver integration for an automation system placed with different devices and systems. This approach can facilitate both an integration in an inside context for WSN and in an integration role between WSN and third-party systems. The remainder of this paper is organized as follow. Section 2 gives you a background information for the discussions in Section 4. Section 3 presents how SWS can be used in WSN. Section 4 also points out some open issues and Section 5 gives the final remarks.

2   Background

2.1   Wireless Sensor Networks

Wireless sensor networks (WSN) are composed of variety number of small, low cost, and dispensable devices called as sensor nodes or just nodes and these nodes can communicate over-the-air to each other using an embedded radio. Nodes have sensing capability and usually are applied to monitor some phenomenon. Nodes also can be deployed very close to some phenomenon or inside it. In order to deliver the collected data to base station, it is common that nodes communicate in an ad-hoc fashion until the data is delivered to the base station. WSN components like nodes, field (deployed area), base station (BS), third-party system, and the relationships among them are shown in Figure 1. In WSN the network size can change from a few to hundreds of thousand nodes.

WSN have some constraints such as low processing capability, limited energy, low transmission rate, and short transmission range [1]. Due to these constraints WSN are application-oriented. WSN can be used for military applications, environment monitoring, factory instrumentation, clean-room monitoring, etc. Currently, the most used operational system in WSN is TinyOS [2]. TinyOS is an event-based OS developed by Berkeley University for its sensor nodes.

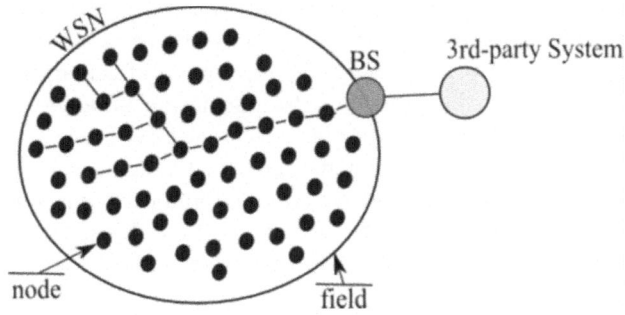

Fig. 1. Components of Wireless Sensor Network

WSN communication can be single or multi hop. So, in order to increase the communication capability and efficiency many protocols were proposed. Each protocol concentrates its effort on a specific service or network layer. Thus different protocols are used for medium access control (B-MAC [3], IEEE 802.15.4 [4]), routing (directed diffusion [5]), data aggregation (Collection Tree Protocol [6]) and dissemination (Location-Aided Flooding [7]), middleware (MANNA [8], TinyDB [9]), and so on. Considering protocols effort, a highlight is the recent approach to adapt TCP/IP stack for WSN. For example, 6loWPAN [10], μIP [11], and a new alliance called IP for Smart Objects (IPSO)[12]. Besides being able to communicate through a wireless communication, WSNs operate on a low-consumption energy [13].

Moreover, after an industry decision on standards about IP adaptation, a large amount of services might emerge and then integration services can be dealt in a easy way. These communication standards could give a step forward in a WSN distinct vendors environment working together properly. For example, electronics from different vendors that are use in ordinary homes could exchange data and then the owner could have a detailed information about energy consumption of those electronics. In addition, semantic web services can have an important role in this new environment.

2.2   Semantic Web Service

A web service can be defined as a piece of software that conforms to a set of open interoperability standards [14] such as Web Service Description Language (WSDL), Simple Object Access Protocol (SOAP), and Universal Description, Discover and Integration (UDDI), for description, messaging protocol and discovering, respectively.

In service-oriented architecture, applications can be accomplished by composing simple services in order to perform a given task. There are several approaches for automating services composition such as Business Process Execution Language (BPEL) [15]. However, most of those approaches give just a syntactic description of the offered service allowing only a predefined service composition. In summary, syntactic services requires human involvement.

On the other hand, Semantic Web Services aim at automating the discovery, composition, invoking and monitoring of Web Services [16] by providing an ontology-based description. The main advantage of semantic web services against syntactic ones is the possibility to associate a parameter to a concept rather than just to a data type. While a syntactic service can only define a parameter isbn of the type String for instance, a semantic service allows the association of this parameter with a International Standard Book Numbers (ISBN) semantic model (e. g. ontology or taxonomy) so that this parameter is able to expect values related to other concepts which has some semantic relationship with ISBN concepts.

In this context, automatic composition consists of dynamic discovery and combine existing services in a registry, given a set of input and output concepts. The matching process should take into account some criteria to determine if two concepts can be considered as the same.

Regarding to dynamic discovery and composition, Paolucci [17] has developed a Matchmaking algorithm that compares advertised services' parameters with request's ones. Kaufer et al. [18] proposed an hybrid service matchmaking based on both logic-based reasoning and matching based on syntactic information retrieval based similarity computation. Sirin et al. [19] proposed a semi-automatic approach for composing semantic web services which means that it is given to the user a set of services whose inputs matches to the output of the previously selected by him. Weise et al. [20] proposed a genetic approach based on evolutionary algorithms.

There are many ways to describe a service in a semantical manner. Some languages like Semantic Annotations for WSDL and XML Schema (SAWSDL) allow description of additional semantics of WSDL components [21]. In summary, they specify how to associate semantic models to a service's description. Another way to describe a service is by modeling the whole service as an ontology. In this sense, OWL-S (Semantic Markup for Web Services) [22] provide a top level ontology of services as shown in Figure 2. This ontology is written in the ontology language OWL (Web Ontology Language) and it is divided into three parts:

- Service Profile: tells "what the service does". It describes the type of information, i. e., concepts involved in a service-seeking process in order to determine whether the service meets its needs.
- Service Model: tells "how the service works". It describes what happen when a service is executing, that is, specifies the details regarding to the order in which and the constraints under which steps that compose the process must be executed.
- Service Grounding: it specifies the details of how an user/agent can access a service, that is, it specifies a communication protocol, message formats, and other service-specific details such as port numbers used in contacting the service.

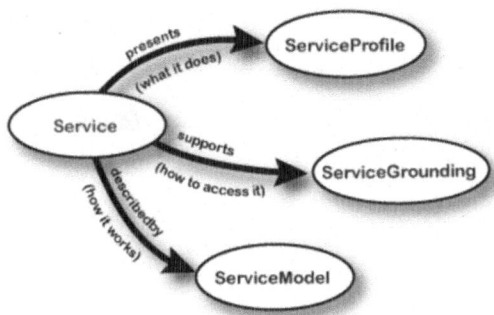

Fig. 2. Top level of the service ontology [22]

3 Applying Semantic Web Services to Wireless Sensor Networks

Sensor Networks is applied in order to provide data from a monitored area using different sources. Those data can be aggregated in order to characterize contexts of different goals. To explore the contexts it could be necessary a system that makes use of data from one or more contexts as well as data processed by other systems. One type of system that can support such processing are Web Services, since they can provide means to facilitate the interoperability with other services. More complex processing can require several service composition schemas, that can include nesting and coordination among services.

As mentioned before in Section 2.1, there are already technologies (e. g. [10,11,12]) that allow to run a minimal webserver over a node, supported by current version of ContikiOS [25]. Then using SWS and WSN together brings us advantages and disadvantages. On the one hand, we can get a flexible system with automatic service composition that could automatic integrate to others system that use same protocol to exchange information. On the other hand, use these technologies together implies an overhead over the traditional WSN. This overhead could be minimized by management an application-dependent trade-off between integration level and what WSN need to perform its goal. Some WSN performance parameter can be network lifetime, throughput, storage size, extra-local processing, and etc.

Thus, in this combined environment, each node can be seen as a service provider and consumer. It can provide a set of services for delivering data and for node's parameters management (network management, operating system, sensing devices, and other application specific values). And it can use some service provided by some node in neighborhood range. Moreover, this service set is dynamic. Nodes can provide or not some service using local or non-local information like remain battery, local data collected, neighborhood data, and so on.

There are at least four different approaches that one can use to deploy SWS in WSN. First approach is when each node has its own service description and implementation, and it does not depend on other non-local services to perform its

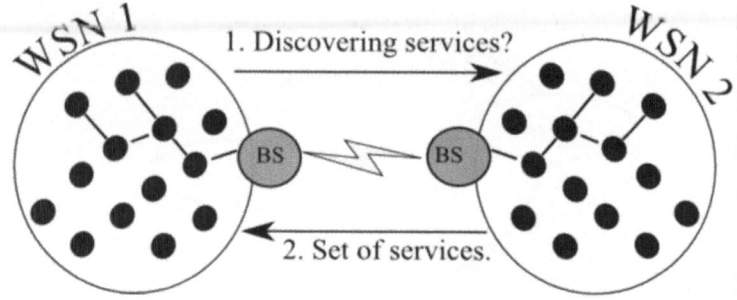

Fig. 3. An example of communication between two WSN with SWS enabled

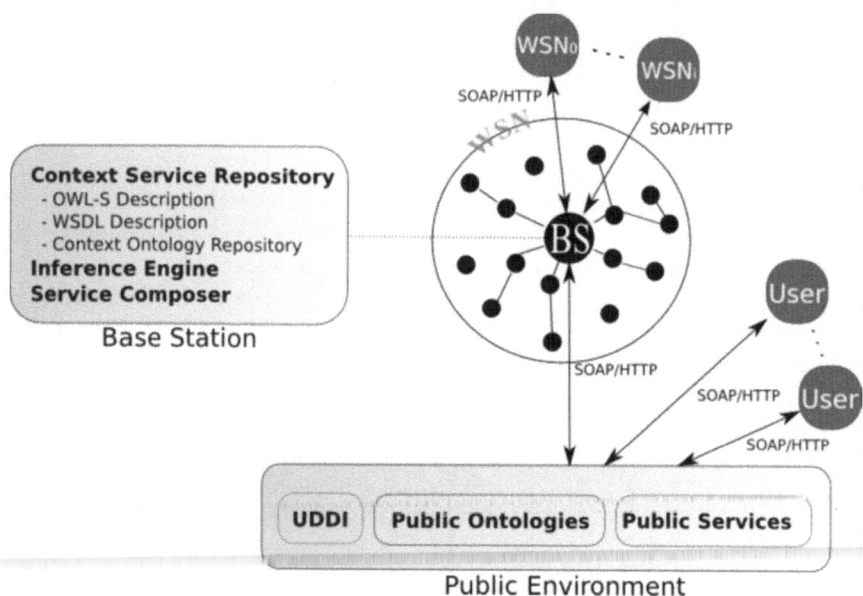

Fig. 4. An SWS option use for Wireless Sensor Network

tasks. The second one is when all related web services tasks are deployed in the base-station and the WSN only provides the data requested. The third approach has its web services deployment in a third-party system. The third-party system receives all the requested data from the base-station. And the last approach is the hybrid one, when all related parts of web services is spread over at least two of these components - nodes, base-station, and third-party system (see Figure 1). In Figure 3 it is shown a communication example between two WSN with SWS enabled in their respective base-station. And in Figure 4 it is shown a more complex scenario where it has WSN exchange information with others one and also it has users using public information through a third-party system.

Therefore, to reach such potential in technology integration a development of ontologies for WSN and SWS is need. These ontologies should meet data and service description. There are few ontologies available, e. g., [23] and [24]. Nevertheless, it is vital a more detailed ontology to help services and applications to create their derived ontology, if necessary, that can preserve the communication capability to each other.

In addition, currently there are different middleware for WSN approaches such as event-based, database-inspired, and others [26]. The middleware aim is to provide network management services, data collection, support for application development, sensing task management, etc. Moreover, they are device, operating system, communication data specific. Then for these middlewares to play an integration and interoperability roles might be difficult and might be need extra effort to achieve them. But to achieve these roles in a given SWS system can play a decisive part. Currently approaches using WSN and SWS are available, e. g., [27,28,29,30,31,32,33].

4 Discussion

Currently, some automation systems make use of wired sensors to perform factory monitoring. Nevertheless, environments that use wired sensors are less flexible for changes than ones that use Wireless Sensors. The main advantage of WSN is the capability to perform wireless communication. For this reason, WSN are more flexible and adjustable. But integration and interoperability in WSN are not consolidated yet.

Many approaches that apply Semantic Web Services to WSN have already been proposed as mentioned in Section 3. However, these approaches are intended to solve problems related only to information gathering and knowledge inference about data collected from a certain environment.

From the point of view of system integration, those approaches are not suitable, since they does not deal with important aspects such as interoperability and description of sensor devices. These are key aspects facilitate the way devices are accessed and managed. Moreover, once devices are described in a semantic manner, it is easy to adapt them according to environment changes. Typical environments are factories, buildings, and homes.

In such environment types, changes can occur on demand or by special needs. Furthermore, these environments are built to serve for different purposes during their lifetime. And for each purpose there is a distinct requirements set. Example of environment changes are re-layout of production line to make a new product or to optimize the process, factory modernization, production lines adjustment for customized products, adaptation to a new layout on a building floor, adjustment for new security and safety rules, owner changes, and so on.

As mentioned before, by using ontology-based service description it is possible to perform dynamic composition of services in way the system is adjusted within a given situation without human-involvement rather than using static and syntactic composition that are widely used in Business Process Management. Furthermore, by interfacing a sensor device, i. e,. by enabling it via (web) services, devices can communicate to each other in a interoperable manner.

Another advantage of applying SWS to WSN is the possibility to hierarchically organize service providers so that fine-grained services (sensor devices) are combined to provide more complex services, in turn are combined with other complex services in order to provide even more complex services.

On the other hand, further studies are needed to evaluate the impact of the overhead brought by high-level protocol message headers in the network performance and lifetime.

4.1 Open Issues

Using SWS with WSN still have issues to be researched, such as:

Standards: This issue plays an essential role to interoperability. It is need standards for communication (MAC, routing, topology control, etc), data representation, service description, service discovery, etc for WSN constraints. Today, we already have some efforts to address this issue, e. g., IEEE 802.15.4 [4], 6loWPAN [10], and ROLL [34].

SWS management inside WSN: SWS deal with service publishing, discovery, composition services, invoking, and monitoring. All these services must have quality of service, fault tolerance, security, dependability, etc then how to fit all of those in such constraint environment like WSN.

Software development process: Currently we have a lack of availability of software development process for WSN applications. Yet, it is need that such process must deal with WSN application development and integration to the whole third-party system.

5 Final Remarks

This paper presents a discussion about applying together wireless sensor networks (WSN) and semantic web services for integration to third-party systems and to others WSNs. This combination might give us a smooth way to deliver integration for an automation system placed with different devices and systems. This approach can facilitate an integration in an inside context for WSN and in an integration role between WSN and third-party systems.

The next step of this research will address the lack of software development process for WSN and the issues related to quality of service for automatic composition of semantic web services. This research will also deal with distinct alternatives to implement the SWS in real WSN with different constraints.

References

1. Akyildiz, I.F., Su, W., Sankarasubramaniam, Y., Cayirci, E.: A Survey on Sensor Networks. IEEE Communications Magazine 40, 102–111 (2002)
2. Hill, J., Szewczyk, R., Woo, A., Hollar, S., Culler, D., Pister, K.: System architecture directions for networked sensors. SIGPLAN Not. 35(11), 93–104 (2000)

3. Polastre, J., Culler, D.: Versatile Low Power Media Access for Wireless Sensor Networks. In: ACM Conf. Embedded Networked Sensor Sys., pp. 95–107 (2004)
4. IEEE Standard for Information technology- Telecommunications and information exchange between systems- Local and metropolitan area networks- Specific requirements Part 15.4: Wireless Medium Access Control (MAC) and Physical Layer (PHY) Specifications for Low-Rate Wireless Personal Area Networks (WPANs), IEEE Std 802.15.4-2006 (Revision of IEEE Std 802.15.4-2003) (2006)
5. Intanagonwiwat, C., Govindan, R., Estrin, D., Heidemann, J., Silva, F.: Directed Diffusion for Wireless Sensor Networking. IEEE/ACM Trans. Networking 11(1), 2–16 (2003)
6. TinyOS Network Protocol Working Group. TEP 123: The Collection Tree Protocol (2008), http://www.tinyos.net/tinyos-2.x/doc/txt/tep123.txt
7. Sabbineni, H., Chakrabarty, K.: Location-Aided Flooding: An Energy-Efficient Data Dissemination Protocol for Wireless Sensor Networks. IEEE Transactions on Computers 54(1), 36–46 (2005)
8. Ruiz, L.B., Nogueira, J.M.S., Loureiro, A.A.F.: MANNA: A Management Architecture for Wireless Sensor Networks. IEEE Communications Magazine 41(2), 116–125 (2003)
9. Madden, S., Franklin, M.J., Hellerstein, J.M., Hong, W.: TinyDB: An Acqusitional Query Processing System for Sensor Networks. In: ACM TODS (2005)
10. Montenegro, G., Kushalnagar, N., Hui, J., Culler, D.: Transmission of IPv6 Packets over IEEE 802.15.4 Networks. IETF RFC 4944 (2007), http://www.ietf.org/rfc/rfc4944.txt
11. Dunkels, A.: Full TCP/IP for 8-bit architectures. In: Proceedings of The First International Conference on Mobile Systems, Applications, and Services (ACM MobiSys 2003), San Francisco, USA (May 2003)
12. IP for Smart Objects (2008), http://www.ipso-alliance.org/
13. Estrin, D., Girod, L., Pottie, G., Srivastava, M.: Instrumenting the World with Wireless Sensor Networks. In: Proceedings of the International Conference on Acoustics, Speech and Signal Processing (May 2001)
14. Pulier, E., Taylor, H.: Understanding Enterprise SOA, Manning, Greenwich (2006)
15. Curbera, F., et al.: Business Process Execution Language for Web Services, v1.0. In: Thatte, S. (ed.) IBM (July 2001), http://www-106.ibm.com/developerworks/webservices/library/ws-bpel
16. Antoniou, G., van Harmelen, F.: A Semantic Web Primer. The MIT Press, Cambridge (2004)
17. Paolucci, M., et al.: Semantic matching of web services capabilities. In: Horrocks, I., Hendler, J. (eds.) ISWC 2002. LNCS, vol. 2342, pp. 333–347. Springer, Heidelberg (2002)
18. Kaufer, F., Klusch, M.: WSMO-MX: A Logic Programming Based Hybrid Service Matchmaker. In: Proceedings of the 4th IEEE European Conference on Web Services (ECOWS 2006). IEEE CS Press, Zurich (2004)
19. Sirin, E., et al.: Filtering and Selecting Semantic Web Services with Interactive Composition Techniques Language for Web. IEEE Intelligent Systems 19, 42–49 (2004)
20. Weise, T., et al.: Different Approaches to Semantic Web Service Composition. In: ICIW 08: Proceedings of the 2008 Third International Conference on Internet and Web Applications and Services, pp. 90–96. IEEE Computer Society, Washington (2008)
21. Farrell, J., Lausen, H. (eds.): Semantic Markup for Web Services (2008), http://www.w3.org/TR/sawsdl/

22. Martin, D., et al.: Semantic Annotations for WSDL and XML Schema (2008),
    http://www.w3.org/Submission/OWL-S/
23. Ota, N., Kramer, W.T.C.: TinyML: Meta-data for Wireless Networks, Technical
    Report, University of California, Berkeley, Computer Science Dept. (2003)
24. Eid, M., Liscano, R., EI Saddik, A.: A Novel Ontology for Sensor Networks Data.
    In: Proceedings of 2006 IEEE International Conference on Computational Intel-
    ligence for Measurement Systems and Applications, La Coruna-Spain, pp. 12–14
    (2006)
25. Dunkels, A., Grnvall, B., Voigt, T.: Contiki - a Lightweight and Flexible Operating
    System for Tiny Networked Sensors. In: Proceedings of the First IEEE Workshop
    on Embedded Networked Sensors (Emnets-I) (2004)
26. Henricksen, K., Robinson, R.: A survey of middleware for sensor networks: state-
    of-the-art and future directions. In: Proceedings of the international workshop on
    Middleware for sensor networks, pp. 60–65 (2006)
27. Whitehouse, K., Zhao, F., Liu, J.: Semantic Streams: A Framework for Composable
    Semantic Interpretation of Sensor Data. In: Römer, K., Karl, H., Mattern, F. (eds.)
    EWSN 2006. LNCS, vol. 3868, pp. 5–20. Springer, Heidelberg (2006)
28. Aberer, K., Hauswirth, M., Salehi, A.: Infrastructure for Data Processing in Large-
    Scale Interconnected Sensor Networks. In: Proceedings of 2007 International Con-
    ference on Mobile Data Management, May 1, pp. 198–205 (2007)
29. Lastra, J.L.M., Delamer, M.: Semantic web services in factory automation: funda-
    mental insights and research roadmap. IEEE Transactions on Industrial Informat-
    ics 2(1), 1–11 (2006)
30. Liu, J., Zhao, F.: Towards semantic services for sensor-rich information systems.
    In: Proceedings of 2nd International Conference on Broadband Networks, vol. 2,
    pp. 967–974 (2005)
31. Gracanin, D., Eltoweissy, M., Wadaa, A., DaSilva, L.A.: A service-centric model
    for wireless sensor networks. IEEE Journal on Selected Areas in Communica-
    tions 23(6), 1159–1166 (2005)
32. Priyantha, B., Kansal, A., Goraczko, M., Zhao, F.: Tiny Web Services for Sensor
    Device Interoperability. In: Proceedings of International Conference on Information
    Processing in Sensor Networks (IPSN 2008), pp. 567–568 (2008)
33. Goodwin, C., Russomanno, D.J.: An Ontology-Based Sensor Network Prototype
    Environment. In: Proceedings of the 5th International Conference on Information
    Processing in Sensor Networks (2006)
34. Routing Over Low power and Lossy networks (2008),
    http://tools.ietf.org/wg/roll/

Beyond Artificial Intelligence toward Engineered Psychology

Stevo Bozinovski[1] and Liljana Bozinovska[2]

[1] Mathematics and Computer Science Department
[2] Biological and Physical Sciences Department
South Carolina State University
300 College street, Orangeburg, SC, 29117
sbozinovski@scsu.edu, lbozinov@scsu.edu

Abstract. This paper addresses the field of Artificial Intelligence, road it went so far and possible road it should go. The paper was invited by the Conference of IT Revolutions 2008, and discusses some issues not emphasized in AI trajectory so far. The recommendations are that the main focus should be personalities rather than programs or agents, that genetic environment should be introduced in reasoning about personalities, and that limbic system should be studied and modeled. Engineered Psychology is proposed as a road to go. Need for basic principles in psychology are discussed and a mathematical equation is proposed as fundamental law of engineered and human psychology.

Keywords: Artificial Intelligence, Consequence-driven Systems theory, personality, Engineered Psychology, limbic system model, fundamental equation of engineered psychology, measuring units in psychology.

1 Introduction

Engineered Psychology is understood as building artificial creatures with human personality features. It considers behavior in a multidimensional *Aristotelian space*, containing at least coordinates of cognition, emotion, and willingness.

In the sequel first we give overview of the Artificial Intelligence development by decades of time starting with 1940-ies ending with current interest toward e-intelligence. Along we present a view on the work done within the ANW research group which is not very well known in the history of AI, in comparison to the work of the PDP group. The overview is then given in terms of phases the AI passed, concluding that now the interest is toward personalities rather than agents. Then we will present our Engineering Psychology approach, starting with the Consequence-driven Systems theory which is origin of the approach and presenting also a model of the brain limbic system. The first result of Engineered Psychology research is the mathematical equation which we call fundamental equation of engineered (and human) psychology. Finally we point out a need for measuring units in engineered and human psychology.

M. Ulieru, P. Palensky, and R. Doursat (Eds.): IT Revolutions 2008, LNICST 11, pp. 171–185, 2009.

2 Artificial Intelligence, So Far

Artificial Intelligence (AI) is contemporary well established area of Computer Science. Universities around the world regularly offer undergraduate course in AI within Computer Science programs. A course usually contains AI history and state of the art presentation of topics related to AI. Several textbooks provide overview of the field [25], [68], [63], [39], [48].

Many authors contributed toward establishing AI as a recognized discipline. The forerunners among them [10], [11], [24], [50], [35], [75], [71], [62], [53], [58], [59], [8], [66] contributed even before the term "Artificial Intelligence" was is use. After 1958 authors [69], [70], [78], [54], [2], [60], [72], [49], [32], [55], pointed toward various approaches toward AI. At the end of this period, due to the influential work [55], one of the approaches toward AI, the neural network approach was understood as restricted, potentially non-promising, which affected funding of that line of research. In 1970-ties, contributors [80], [28], [1], [26], [37], [81], [9], [76], [51], continued showing interest in various aspects of representing and manifesting AI. Particular belief was shown that special programming languages such as LISP and PROLOG should be used to foster development of AI programs.

The decade of 1980-ties is arguably the revolutionary step toward so called "new AI" or "embodied AI" or "behavior based AI" as distinction to the previous "disembodied AI". The works such as [5], [14], [6], [33], [38], [36], [22], [23], [67], [56], [73], [74], [77], [31] contributed toward importance of agent-environment interaction. Parallel programming was introduced as a way of representing that interaction [13] as well as the need of sensors and actuators in an agent-environment interface [22]. In this period the neural network research was revived. We would emphasize the work of two organized group working in that direction: the Adaptive Networks Group (ANW) and the Parallel Distributed Processing (PDP) group. In stories about AI the work of PDP group is well known: they published an important volume on Parallel Distributed Processing (PDP) paradigm and explicitly rehabilitated the neural networks fields as very promising one within AI. The ANW group never published a volume together as a group so work as a group was not recognized within AI community. In the next section we will give a review of that work in the early 1980-ies.

During 1990-ties single important event overshadowed others: On May 11, 1997, computer program Big Blue beat the human chess champion! The first grand challenge of AI was no more! Consequently, a new grand challenge for AI was set up: Is it possible that a robot team in (European) football of 11 players would beat a corresponding human team [3]? And variation of that: Is it possible that a trained (rather than programmed) team [16] would achieve the same? Among other events in 1990-ies, it was pointed out that AI is not artificial by itself, it should be considered a natural intelligence in artificial creatures [4]. Increased interest has been shown toward emotions and its relation to intelligence ([41], [65], [42], [61], [16]).

In the 2000-ies early workers of AI such as [27], [56] also devoted books related to emotion. Emotion became topic of graduate Computer Science seminars [82]. Motivation and emotion became modules of control architectures. But most important phenomenon is the spread of AI among high school through robotics competitions. Robotics and related AI issues become widespread challenge for talents all over the world. Also

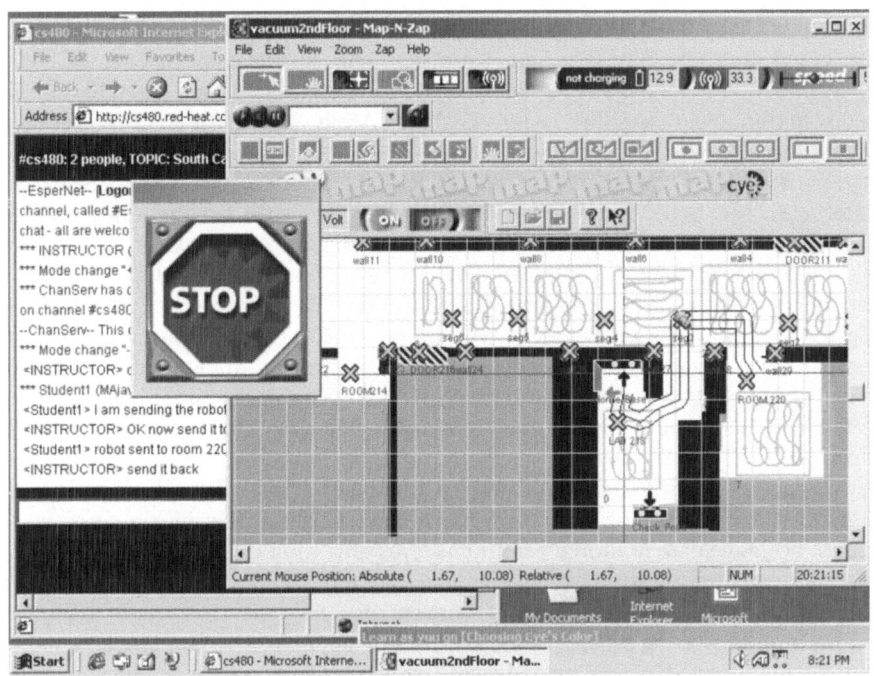

Fig. 1. A e-controlled mobile robot, cognitive model of its world

Internet and e-robotics (e-control of robots) is widely used. Fig. 1 shows our work with a commercially available e-controllable cye-type robot.

The robot is user programmable to have internal model of its environment, but is also able to learn new obstacles. The environment shown here is the AI/Robotics/Biocybernetis lab at South Carolina State University with corridor and other rooms. In particular Fig. 1 shows a session of our CS480 Introduction to Robotics class in distance learning using e-robotics [18]. Students are able using Internet to send the robot to a particular environment region to perform a vacuum cleaning task. In this particular e-robotics experiment the robot is controlled from a classroom in Europe. Fig. 1 shows how robot is executing a trajectory which it computes itself to avoid obstacles (dark areas) in order to reach its vacuum cleaning region. While the robot is moving the STOP sign is present so the robot can be stopped by click of a mouse. Left side of Fig. 1 shows the virtual classroom in which the instructor in USA communicates with students in Europe in real time. The experiment includes also a camera (not shows in Fig. 1) mounted on the robot so the students can scan the environment while the robot moves.

2.1 The ANW Group in Early 1980-ies

In early 1980-ties, possibly due to belief that artificial neural networks line of research is not promising one, in the USA there were not many organized groups working on the subject. In fact we know of only one: the Adaptive Networks (ANW)

group, funded by Wright-Patterson Air Force Base (WPAFB) from Dayton, Ohio. The group was assembled by the Computer and Information Science (COINS) department of the University of Massachusetts at Amherst. Nico Spinelli was the group leader, who together with Michael Arbib led a Center for Systems Neuroscience within the department. During 1980-81 the group included the postdoc Andrew Barto, and the graduate students Richard Sutton, Charles Anderson, Jack Porterfield, Ted Selker, and Stevo Bozinovski. The main driving force was the WPAFB Program Officer, Harry Klopf, and his motto *"goal-seeking systems from goal-seeking components"* [40]. The neural networks paradigm well corresponds to that motto and the group pursued that direction. The group mostly oriented its research toward agent-environment type of research such as agents moving in an environment and agents moving objects in an environment. Maintaining some effort on classical conditioning and patters classification, the group focused on the concept of reinforcement and reinforcement learning ([53], [52], [64], [79]). The initial important result in this direction was the Associative Search Network (ASN), an agent that was able to learn to navigate toward a goal place using landmarks [5]. Next, the group took a challenge of *delayed reinforcement* learning, learning in cases where there is no immediate reinforcement (punish/reward) from the environment, rather it comes after several steps. That challenge led to introduction of emotions and genetics in neural nets and we present it here in more details.

Two instances of the problem were considered: the maze learning problem and the learning for inverted pendulum balancing problem. Two neural architectures were proposed for solving the problem: the Actor/Critic architecture, and the Crossbar Adaptive Array (CAA) architecture (Fig. 2). It is interesting to analyze the difference between the two architectures. The obvious difference is that A/C architecture needs two identical memory structures, V and W, to compute the internal reinforcement $r^\wedge$, and the action, while CAA architecture uses only one memory structure, W, the same size as one of the A/C memory structures. The most important difference however is in the *design philosophy*: In contrast to A/C architecture, *CAA architecture does not use any external reinforcement r*; it only uses the current situation X as input. A state value is used as secondary reinforcement concept, while genetically predefined state value, rather than an immediate reinforcement, is used as primary reinforcement. The CAA architecture *introduces the concept of state evaluation* and connects it to *the concept of feeling*. Although both architectures were designed to solve the same basic problem, the maze learning problem, the challenges of the design were different. The A/C architecture effort was challenged by the mazes from animal learning experiments, where there are many states and there is only one rewarding state (food). The CAA architecture effort from the start was challenged by the mazes defined in the computer game of Dungeons and Dragons, where there is one goal state, but many reward and punishment states along the way. So, from the very beginning CAA effort adopted the concept of dealing with *pleasant and unpleasant states*, feelings and emotions.

The CAA approach proved more efficient: it was the only architecture that presented solutions to both instances of the delayed reinforcement learning problem in 1981 ([12], [13], [14], [6], [15]). The original CAA idea of having *one memory structure for crossbar computation of both state evaluations and action evaluations* was later also implemented in reinforcement learning architectures such as Q-learning system [7], [77] and its successor architectures. The CAA memory was named Q-table.

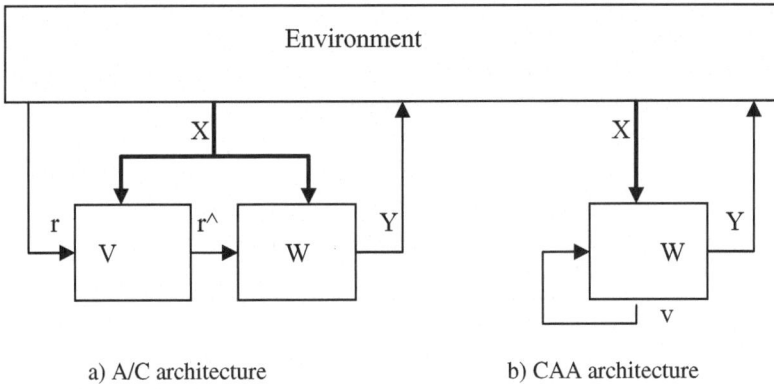

a) A/C architecture   b) CAA architecture

Fig. 2. ANW neural architectures for solving delayed reinforcement learning problem

The CAA approach introduced neural learning systems that can *learn without external reinforcement.* It proposes a *paradigm shift* in learning theories, from the concept of reinforcement (reward, punishment, payoff, …) to the concept of *state as a consequence* and *feeling as a state evaluation.* Including concepts like feelings and emotions, the CAA approach introduced emotive abilities in cognitive neural based agents.

2.2 Phases of AI Development: Programs, Agents, Personalities

Reviewing trajectory of the AI research so far we can distinguish three phases. The initial, zero-phase would be the cybernetics phase, where researchers were mostly interested in control structures, adaptation, and pattern recognition.

The first phase is so called representational (or disembodied) phase where researchers were mostly interested in representations of environments in internal knowledge structures of agents. The principal metaphor was information processing and AI programs, including special programming languages. Fig. 3 shows the principal postulates of AI understanding:

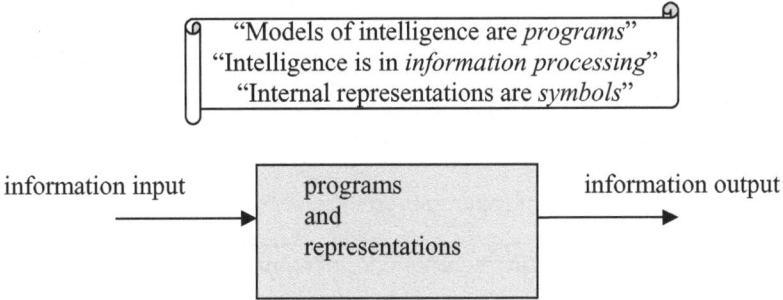

Fig. 3. Understanding artificial intelligence, representational phase

Second phase emphasizes behavior as representation of intelligence. Fig. 4 shows the behavioral metaphor and its main postulates of AI understanding. This phase is also known the embodied phase of AI. In order to sense the reality an AI agent needs sensors, and to wear sensors an agent needs a body.

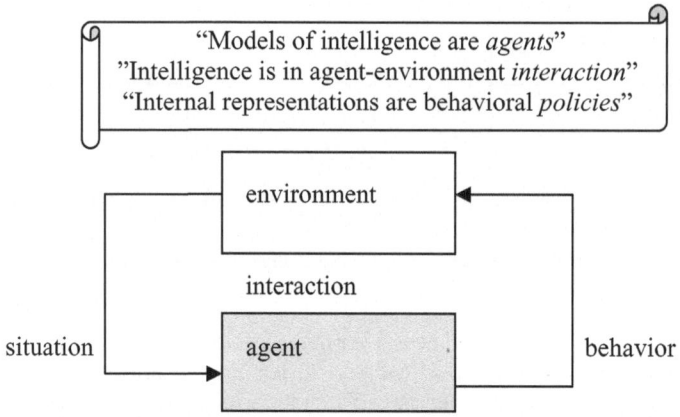

Fig. 4. Understanding artificial intelligence, behavioral (embodied) phase

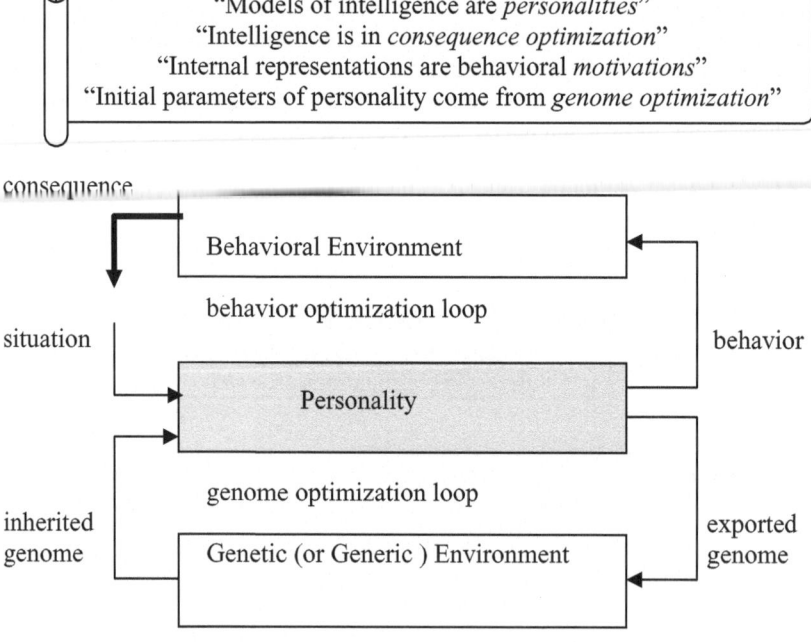

Fig. 5. Understanding Artificial Intelligence, personalities phase

The third phase we propose is the personality phase. Fig. 5 shows the personality metaphor and its principal postulates. The concept of personality assumes comparison of two behaviors and features that distinct one personality from another one. It is assumed that a personality cannot be considered separated neither from its behavioral environment nor from its genetic (generic) environment. Genome optimization loop (evolution) ensures optimization of the initial parameters of personality and its connection to the changing behavioral environment.

3   Engineered Psychology Rather Than Artificial Intelligence

Personality can be understood as *expectancy of a consistent pattern of attitudes and behavior*. All people exhibit recognizable individual actions that serve to identify them. Classical problems in personality include: Where do those individual characteristics come from? Are they truly unique or just particular combination of characteristics all people possess? Are they learned, inherited, or both? Can personality be altered, and if so, how? [46]

For many years Artificial Intelligence research attention was primarily given to the *concept of intelligence,* not necessarily an embodied one. In recent years, growing interest has been in building agents that will interact with humans, and will exhibit features of personality. The effort can be described as understanding the *concept of personality* and building an *artificial personality*, with its features like emotions and motivations, among others. In the sequel we address some of the challenges of artificial personality research within our theory of Consequence-driven Systems.

3.1   Consequence-Driven Systems Theory

Consequence-driven Systems theory [14], [17], [20], [21] is an attempt to understand and build an *agent (animat) personality*. It tries to *find an architecture* that will ground the notions such as motivation, emotion, learning, disposition, anticipation, curiosity, confidence, and behavior, among other notions usually present in a discussion about an agent personality. Searching through the literature we found relevant works that can be considered as grounds on which our work is a continuation. In particular we found a highly relevant connection with the works [43], [44], [45], [47]. Among several related efforts in contemporary research on emotion-based architectures, we distinguish the work [29], [30] by its basic concepts, architecture engineering, and realization aspects that resulted in developed an emotion learning architecture that has been implemented successfully in simulated as well as real robots. Starting from concepts such as feelings and hormonal system, the approach has similarity with our approach in issues like having innate emotions that will define goals and learning emotional associations between states and actions that will determine the agent's future decisions.

In the sequel we will briefly review the Consequence Driven Systems theory and its features.

3.1.1 Main Concepts of the Theory

The grounding postulates of the Consequence Driven Systems theory are: 1) there are three environments, 2) there are three tenses, 3) the neural system computes simultaneously both behaviors and emotions, from its memory 4) emotions are computed as state evaluations, 5) motivations are learned as behavior evaluations and 6) what is obtained should correlate to the brain limbic system.

Three environments. The theory assumes that an agent should always be considered as a three-environment system. The agent expresses itself in its *behavioral environment,* where it behaves. It has its own *internal environment* where it synthesizes its behavior. It also has access to a *genetic environment* where from it receives its initial conditions for existing in its behavioral environment. The genetic and the behavioral environment are related: The initial knowledge transferred through the imported *species (or personality) genome* properly reflects, *in the value system of the species,* the dangerous situations for the species in the behavioral environment. It is assumed that all the agents import some genomes at the time of their creation. Some of the agents will be able to export their genome.

Three tenses. The theory emphasizes that *an agent should be able to understand the temporal concepts,* like past, present, *and the future,* in order to be able to self-organize in an environment. The past tense is associated with the evaluation of its previous performance, the present tense is associated with the concept of emotion that the agent computes toward the current situation (or the current state), and the *future* with the *moral (self-advice)* about a behavior that in future should be associated with a situation.   A *Generic Architecture for Learning Agents* (GALA architecture) is proposed within the theory (Fig. 6).

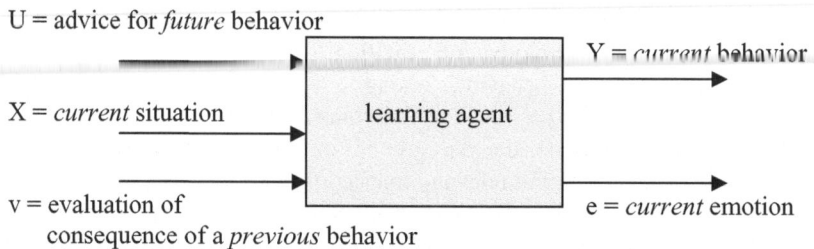

U = advice for *future* behavior

Y = *current* behavior

X = *current* situation

learning agent

v = evaluation of
consequence of a *previous* behavior

e = *current* emotion

Fig. 6. The GALA architecture

Note that GALA is a genuine generic, black-box-only architecture: only the inputs and outputs are specified. Yet it is very specific the way inputs and outputs are defined. The GALA architecture is a reconfigurable architecture: The derivation of various types of learning architectures from the GALA architecture is described elsewhere [14], [19].

Crossbar-adaptive array architecture. The CAA architecture (Fig. 7) is derived as an emotion learning agent from the GALA architecture.

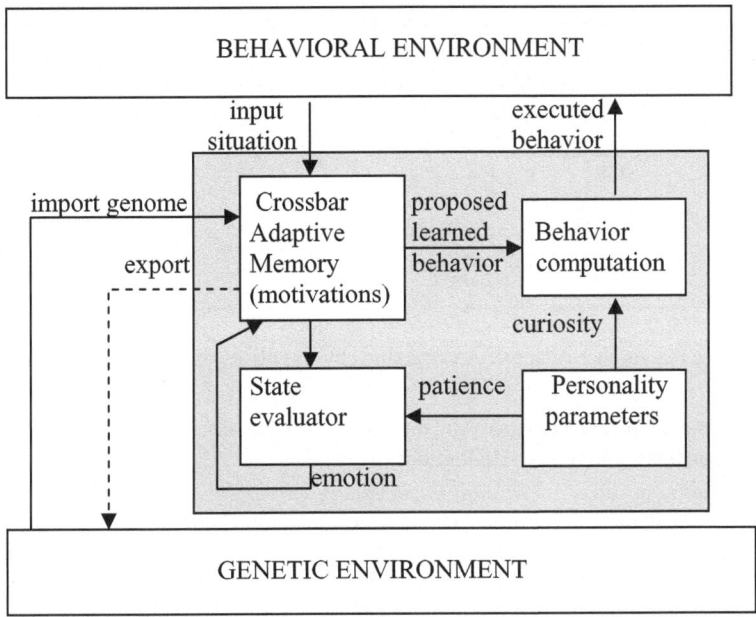

Fig. 7. Crossbar-adaptive Array (CAA) architecture

The CAA architecture is a personality architecture capable of learning using *backpropagated emotions*. In a crossbar fashion, the CAA architecture computes both state evaluations (emotions) and behavior evaluations (motivations). It contains three basic modules: crossbar-learning memory, state evaluator, and behavior selector. In its basic behavioral cycle, the CAA architecture firstly computes the emotion of being in the current state. Then, using a feedback loop, it computes the possibility of choosing again, in a next time, the behavior to which the current situation is the consequence. The state evaluation module computes the global emotional state of the agent and *broadcasts it* (e.g. by way of a neuro-hormonal signal) to the crossbar-learning memory. The behavior computation module using some kind of behavior-algebra initially performs a *curiosity driven*, default behavior, but gradually that behavior is replaced by a learned behavior. The forth module defines specific *personality parameters* of a particular agent, such as *curiosity*, *tolerance* threshold (*patience*), etc. Indeed CAA includes models of those personality paramets.

Correlation to the limbic system. A possible model of the limbic system that correlates to the CAA architecture is given in Fig. 8.

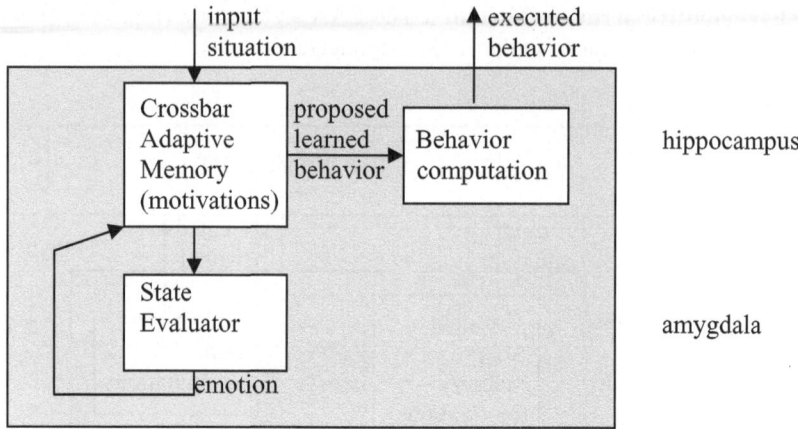

input
situation

executed
behavior

Crossbar
Adaptive
Memory
(motivations)

proposed
learned
behavior

Behavior
computation

hippocampus

State
Evaluator

amygdala

emotion

Fig. 8. Crossbar Adaptive Array architecture as model of the limbic system

The Crossbar Adaptive Array as model of the limbic system contains both the hippocampus level and amygdale level of processing.

Emotion. Is evaluation of a situation (or state). Usually that situation is observed as a consequence of a previous behavior.

Motivation. The motivation for a behavior is the anticipation of a future consequence of that behavior; the emotion is an evaluation system for computing the value of that consequence. The motivational system can be understood as a priority system for executing behaviors. Motivations are computed from the values of their (anticipated) consequence states. The Consequence-driven System theory introduced the *principle of learning and remembering only motivations (behavior evaluations), not the state evaluations.* There is no need to store the state values, since they can be computed from the (current) behavior values. The approach of storing action values rather than state values was also emphasized by Watkins [77], who discovered the relationship between Reinforcement Learning and Dynamic Programming [8]. There are different approaches of what an agent learns in an environment. It can learn the whole graph, like a *cognitive map*, or it can learn only a *policy* (the set of states and behaviors associated to those states). In case of policy learning the actual map (interconnection network between the states) is provided by the environment and is not stored within the agent.

3.2 Fundamental Law of Engineered (and Human) Psychology

Since beginning of philosophy and psychology, behavior, motivation for a behavior, and emotions related to a situation have been of interest for science. Mathematical models were proposed for those concepts. However it was only recently [19], [20] when the first mathematical model was proposed that *explicitly related* motivation, emotion, behavior and personality. Fig. 9 shows the approach toward the modeling. The approach is general and applies to both human and engineered psychology.

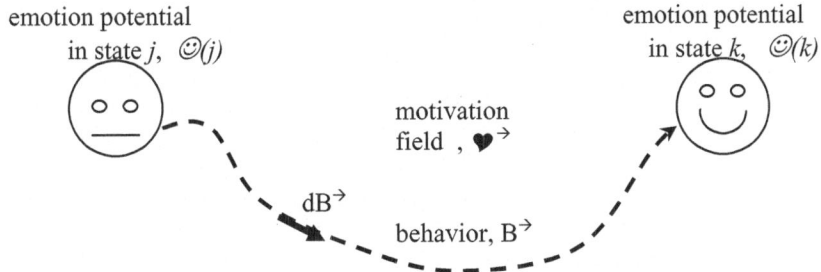

emotion potential
in state *j*, ☺*(j)*

emotion potential
in state *k*, ☺*(k)*

motivation
field , ❤$^\rightarrow$

dB$^\rightarrow$

behavior, B$^\rightarrow$

Fig. 9. Emotional space and motivational gradient

As Figure 9 shows, it is assumed that each state *j* of a system (e.g. animat) is assigned an *emotion potential* ☺*(j)*. The emotion potential space defines by itself a vector field denoted as *motivational field*, ❤$^\rightarrow$. In such a way each point in the space is assigned a *motivation gradient* toward the highest emotional value (ultimate goal) at that region of the space.

$$☺ = p \int_{j}^{k} ❤^\rightarrow dB^\rightarrow .$$ (1)

In such a setup, a behavior $B^\rightarrow$ (that is motivated by the motivation field ❤$^\rightarrow$ induced by the emotional potential ☺) obeys the relation (1). Equation (1) connects the emotion potential in a current state *j*, an *anticipated emotion potential* in a next state *k*, the motivation of moving toward state *k*, and behavior toward the state *k*. Here $dB^\rightarrow$ is a differential segment of a behavior, an *action,* which sometimes might not be in the same direction as the motivational field, for example due to some obstacles (resistances) along the way. The constant *p* allows for some personality features of a particular agent. In other words the equation (1) is a mathematical way of saying that the *happiness is to be earned.* We propose this equation as a *fundamental equation of engineered (and human) psychology.*

Related to the equation (1) is an algebraic *motivation learning function* which can be written as

$$❤(b|s) \mathrel{+}= (☺(s') - ☺(b|s))$$ (2)

where += is the update operator we borrowed from object oriented languages for use in the simulation experiment described further in this text. The equation (2) simply states that *the motivation update is proportional to the satisfaction expected.* In more detail, it states that the motivation ❤*(b|s)* of performing behavior *b* in *s* depends on the personally anticipated emotion ☺*(s')* in the *consequent situation s'* but also on the personally *anticipated effort* ☺*(b|s)* needed for executing behavior *b* in situation *s.* We introduce the concept of *willingness to perform a behavior* as a function of (☺*(s')* - ☺*(b|s)*).

Simulation work that shows the validity of this theory is given in [20].

3.3 Basic Principles and Measuring Units

Comparing to physics, psychology is still just an empirical science. There are no basic principle equations like F = ma, and U = IR, and there is no system of units in for emotion and other concepts, in comparison to physics with units like kilogram for mass and Newton for force. An attempt for establishing basic principles might be indeed the equation (1). It relates behavior emotion and motivation, which means that at least one of those psychological concepts does not need a special measuring unit, it can be inferred from the other two. In [20] behavior is left to be such a concept, while emotion is measured in (… 2☻, ☻, ☺, ☺, 2☺, …) units and motivation in (…,-2♥, -♥, 0♥, ♥, 2♥, …) units. Any names can be given to those units and for now we call them Dze' and Daron respectively (names are from old macedonian mythology). So in engineered psychology we may talk in terms of motivation of 5 Darons and emotion of -3 Dze's. Let us note that Smiley is widely used name for the ☺-emoticon. Other names might also be proposed.

4 Conclusion

This work suggests that it is time review the undergraduate courses in AI, with direction toward Engineered Psychology. Maybe is time to replace the topic about search techniques which are now standard part of theory of algorithms, and go toward topics beyond intelligence and cognition, by including other features of a personality.

A result of Engineered Psychology approach is the first mathematical model that explicitly relates motivation, emotion, behavior, and personality. To the best of our knowledge such a model has not been proposed nether in psychology nor in Artificial Intelligence research. By building psychology in artificial creatures we believe we will better understand psychology in general. A step has been made toward the search for basic principles and measuring units in both engineered and human psychology.

References

1. Anderson, J., Bower, G.: Human Associative Memory. Erlbaum, Mahwah (1973)
2. Arbib, M.A.: Brains, Machines, and Mathematics. McGraw Hill, New York (1964)
3. Asada, M., Kitano, H., Noda, I., Veloso, M.: Robocup: Today and Tomorrow – What we Learned. Artificial Intelligence 110, 193–214 (1999)
4. Balkenius, C.: Natural Intelligence in Artificial Creatures. Lund University Press (1995)
5. Barto, A., Sutton, R.: Landmark Learning: An Illustration of Associative Search. Biological Cybernetics 42, 1–8 (1981)
6. Barto, A., Sutton, R., Anderson, C.: Neuronlike Elements that can Solve Difficult Learning Control Problems. IEEE Transactions on Systems, Man, and Cybernetics 13, 834–846 (1983)
7. Barto, A., Sutton, R., Watkins, C.: Learning with Sequential Decision Making. In: Gabriel, M., Moore, J. (eds.) Learning and Computational Neuroscience: Fundamentals of Adaptive Networks, pp. 539–602. MIT Press, Cambridge (1990)
8. Bellman, R.: Dynamic Programming. Princeton University Press, Princeton (1957)
9. Bledsoe, W.: Non-resolution Theorem Proving. Artificial Intelligence 9 (1977)

10. Boole, J.: An Investigation of the Laws of Thought. Walton and Maberly (1854)
11. Bouton, A.: A Game with a Complete Mathematical Theory. Annals of Mathematics Princeton 3, 35–39 (1901/1902)
12. Bozinovski, S.: A Self-learning System using Secondary Reinforcement. ANW Report, November 25, 1981. COINS. University of Massachusetts at Amherst (1981)
13. Bozinovski, S.: Inverted Pendulum Learning Control. ANW Memo. December 10, 1981. COINS. University of Massachusetts, Amherst (1981)
14. Bozinovski, S.: A Self-learning System Using Secondary Reinforcement. In: Trappl, R. (ed.) Cybernetics and Systems Research, pp. 397–402. North-Holland, Amsterdam (1982)
15. Bozinovski, S.: Crossbar Adaptive Array: The First Connectionist Network that Solved the Delayed Reinforcement Learning Problem. In: Dobnikar, A., Steele, N., Pearson, D., Alberts, R. (eds.) Artificial Neural Networks and Genetic Algorithms, pp. 320–325. Springer, Heidelberg (1999)
16. Bozinovski, S., Jaeger, H., Schoell, P.: Engineering Goalkeeper Behavior using Emotion Learning Method. In: Sablatnoeg, S., Enderle, S. (eds.) Proc. Workshop on RoboCup, KI 1999, pp. 48–56. Deutsche Jarhestagung fuer Kuenstliche Intelligenz, Bonn (1999)
17. Bozinovski, S., Bozinovska, L.: Self-learning Agents: A Connectionist Theory of Emotion Based on Crossbar Value Judgement. Journal of Cybernetics and Systems 32, 637–669 (2001)
18. Bozinovski, S., Jovancevski, G., Ackovska, N.: Distributed Interactive Robotics Classroom. Optoelectronics Information-Power Technologies 2(4), 5–9 (2002)
19. Bozinovski, S.: Motivation and Emotion in Anticipatory Behavior of Consequence Driven Systems. In: Butz, M., Sigaud, O., Gerard, P. (eds.) Proc. Workshop on Adaptive Behavior in Anticipatory Learning Systems, Edinburgh, pp. 100–119 (2002)
20. Bozinovski, S.: Anticipation Driven Artificial Personality. Building on Lewin and Loehlin. In: Butz, M.V., Sigaud, O., Gérard, P. (eds.) Anticipatory Behavior in Adaptive Learning Systems. LNCS, vol. 2684, pp. 133–150. Springer, Heidelberg (2003)
21. Bozinovski, S., Bozinovska, L.: Evolution of a Cognitive Architecture for Emotional Learning from a Modulon-structured Genome. Journal of Mind and Behavior 29(1-2), 195–216 (2008)
22. Braitenberg, V.: Vehicles: Experiments in Synthetic Psychology. MIT Press, Cambridge (1986)
23. Brooks, R.: A Robust Layered Control for a Mobile Robot. IEEE Journal of Robotics and Automation 2, 14–23 (1986)
24. Church, A.: The Calculi of Lambda Conversion. Annals of Mathematical Studies, vol. 6. Princeton Univesity Press, Princeton (1941)
25. Crevier, D.: AI: The Tumultuous History of the Search for Artificial Intelligence. Basic Books, New York (1993)
26. Duda, R., Hart, P.: Pattern Classification and Scene Analysis. John Wiley, Chichester (1973)
27. Fellous, J.-M., Arbib, M. (eds.): Who Needs Emotions? Oxford University Press, Oxford (2005)
28. Fikes, R., Hart, P., Nilsson, N.: Learning and Executing Generalized Robot Plans. Artificial Intelligence 3(4), 251–288 (1972)
29. Gadanho, S.: Reinforcement Learning in Autonomous Robots: An Empirical Investigation of the Role of Emotions. PhD Thesis. University of Edinburgh (1999)
30. Gadanho, S.: Learning Behavior-selection by Emotions and Cognition in a Multi-goal Robot Task. Journal of Machine Learning Research 4, 385–412 (2003)

31. Goldberg, D.: Genetic Algorithms in Search, Optimization, and Machine Learning. Addison-Wesley, Reading (1989)
32. Green, C.: Theorem-proving by Resolution as a Basis for Question-answering System. In: Michie, D., Meltzer, B. (eds.) Machine Intelligence, vol. 4, pp. 183–205 (1969)
33. Grossberg, S.: Studies of Mind and Brain: Neural Processing of Learning, Perception, Development, and Motor Control. Reidel Press, Boston (1982)
34. Haugeland, J.: Artificial Intelligence: The very Idea. MIT Press, Cambridge (1985)
35. Hebb, D.: The Organization of Behavior. Wiley, Chichester (1949)
36. Hillis, D.: The Connection Machine. MIT Press, Cambridge (1985)
37. Holland, J.: Adaptation in Natural and Artificial Systems. University of Michigan Press, Ann Arbor (1975)
38. Hopfield, J.: Neural Networks and Physical Systems with Emergent Collective Computation Abilities. In: Proc. National Academy of Science, vol. 79, pp. 2554–2558 (1984)
39. Jones, T.: Artificial Intelligence: A Systems Approach. Infinity Science Press (2008)
40. Klopf, H.: Goal-seeking Systems from Goal-seeking Components: Implementation to AI. The Cognition and Brain Theory Newsletter 3 (1979)
41. LeDoux, J.: Emotion and the Limbic System Concept. Concepts in Neuroscience 2, 169–199 (1991)
42. LeDoux, J.: The Emotional Brain. Simon and Schuster (1996)
43. Lewin, K.: Dynamic Theory of Personality. McGraw Hill, New York (1935)
44. Lewin, K.: Principles of Topological Psychology. McGraw Hill, New York (1936)
45. Lewin, K.: Field Theory in Social Science. Harper and Row (1951)
46. Liebert, R., Spiegler, M.: Personality. The Dorsey Press (1974)
47. Loehlin, J.: Computer Models of Personality. Random House (1968)
48. Luger, G.: Artificial Intelligence: Structures and Strategies for Complex Problem Solving. Addison Wesley, Reading (2009)
49. McCarthy, J.: Programs with Common Sense. In: Minsky (ed.) Semantic Information Processing, pp. 403–418 (1968)
50. McCulloch, W., Pitts, W.: A Logical Calculus of the Ideas Immanent in Nervous Activity. Bulletin of Mathematical Biophysics 5, 115–133 (1943)
51. McDermot, D.: Planning and Acting. Cognitive Science 2, 71–109 (1978)
52. Mendel, J.: Reinforcement Learning Control and Pattern Recognition Systems. In: Mendel, J., Fu, K. (eds.) Adaptive, Learning, and Pattern Recognition Systems: Theory and Applications, pp. 287–318. Academic Press, London (1970)
53. Minsky, M.: Theory of Neural-analog Reinforcement Systems and its Application to the Brain-model System. Ph.D. Thesis, Princeton University (1954)
54. Minsky, M.: Steps toward Artificial Intelligence. In: Proceedings IRE, vol. 49, pp. 8–30 (1961)
55. Minsky, M., Papert, S.: Perceptrons: An Introduction to Computational Geometry. MIT Press, Cambridge (1969)
56. Minsky, M.: The Society of Mind. Simon and Schuster (1986)
57. Minsky, M.: The Emotion Machine. Simon and Schuster (2006)
58. Newel., A.: The Chess Machine: An Example of Dealing with Complex Tasks of Adaptation. In: Proc. Western Joint Computer Conference, pp. 101–108 (1955)
59. Newel, A., Shaw, J., Simon, H.: Empirical Explorations of the Logic Theory Machine: A Case Study of Heuristics. In: Proceedings Western Joint Computer Conference, vol. 15, pp. 218–239 (1957)
60. Nilsson, N.: Learning Machines. McGraw-Hill, New York (1965)

61. Petta, P., Trappl, R.: Personalities for Synthetic Actors: Current Issues and Some Perspectives. In: Petta, P., Trappl, R. (eds.) Creating Personalities for Synthetic Actors. LNCS, vol. 1195, pp. 209–218. Springer, Heidelberg (1997)
62. Piaget, J.: The Construction of Reality in the Child. Basic Books, New York (1954)
63. Pfeifer, R., Scheier, C.: Understanding Intelligence. The MIT Press, Cambridge (2000)
64. Rescorla, R., Wagner, A.: A Theory of Pavlovean Conditioning: Variations in the Effectiveness of Reinforcement and Non-reinforcement. In: Black, A., Procasy, W. (eds.) Classical Conditioning II. Current Research and Theory, pp. 64–99. Appleton-Century-Croffts (1972)
65. Rolls, E.: A Theory of Emotion and Consciousness, and its Application in Understanding the Neural Basis of Emotion. In: Gazzaniga, M. (ed.) The Cognitive Neuroscience, pp. 1091–1106. The MIT Press, Cambridge (1995)
66. Rosenblatt, F.: The Perceptron: A Probabilistic Model for Information Storage and Organization in the Brain. Psychological Review 65, 386–408 (1958)
67. Rumelhart, D.: McClelland and the PDP Research Group. In: Parallel Distributed Processing. MIT Press, Cambridge (1986)
68. Russel, S., Norvig, P.: Artificial Intelligence: A Modern Approach. Prentice-Hall, Englewood Cliffs (2002)
69. Samuel, A.: Some Studies of Machine Learning Using the Game of Checkers. IBM Journal of Research and Development 3, 211–229 (1959)
70. Selfridge, O.: Pandemonium: A Paradigm of Learning. In: Symposium of the Mechanization of Thought Process, pp. 513–526. HMSO (1959)
71. Shannon, C.: Programming a Computer for Playing Chess. Philosophical Magazine 41 (1950)
72. Simon, H.: Motivational and Emotional Controls of Cognition. Psychological Review 74, 29–39 (1967)
73. Staugaard, A.: Robotics and AI: An Introduction to Applied machine Intelligence. Prentice-Hall, Englewood Cliffs (1987)
74. Sutton, R.: Learning to Predict by the Method of Temporal Difference. Machine Learning 3, 9–44 (1988)
75. Turing, A.: Computer Machinery and Intelligence. Mind 59, 433–460 (1950)
76. Warren, D., Pereira, L., Pereira, F.: PROLOG – the Language and its Implementation Compared with LISP. Proc. Symp. on AI and Programming Languages. SIGPLAN Notices 12(8) (1977)
77. Watkins, C.: Learning from Delayed Rewards. Ph.D. Thesis. King's College, Cambridge (1989)
78. Widrow, B., Hoff, M.: Adaptive Switching Circuits. IRE WESTCON Convention Record, pp. 96–104 (1960)
79. Widrow, B., Gupta, N., Maitra, S.: Punish/Reward: Learning with a Critic in Adaptive Threshold Systems. IEEE Trans. Systems, Man, and Cybernetics 3(5), 455–465 (1973)
80. Winograd, T.: Understanding Natural Language. Academic Press, London (1972)
81. Wirth, N.: Algorithms + Data Structures = Programs. Prentice Hall, Englewood Cliffs (1976)
82. Siegelmann, H.: Computational Models of Emotion. CS691I Seminar, University of Massachusetts at Amherst (2007),
http://www.cs.umass.edu/~molsen/emotion

Communication in Change – Voice over IP in Safety and Security Critical Communication Networks

Heimo Zeilinger, Berndt Sevcik, Thomas Turek, and Gerhard Zucker

Vienna University of Technology, Institute of Computer Technology,
Gusshausstr. 27-29, 1040 Vienna, Austria
{zeilinger,sevcik,turek,zucker}@ict.tuwien.ac.at

Abstract. During the last decade communication technology has changed rapidly. Due to its decreasing costs and rising expansion, IP (Internet Protocol) technology has found its way to areas that have long been the domain of public-switched telephone networks (PSTN). Voice over IP (VoIP) applications are widely used not only for phone calls or common Internet conferences, but also tend to be used for safety critical communication applications. Hence security and safety topics arise, which pose new challenges in this area of research. The authors are convinced that new issues on the network layer as well as on the application layer require detailed analysis. Hence this paper gives an overview on latest developments in this area, and states the authors' view on this topic. Thereby safety and security issues are faced from different abstraction layers. On the one hand the network layer and on the other hand the application layer focusing on middleware systems in the area of service oriented architectures (SOAs).

Keywords: VoIP, communication networks, safety, security.

1 Introduction

Within the last decade VoIP benefited from the massive expansion of IP networks and related decrease in acquisition and operation costs. Today a trend towards the use in public safety communication networks can be identified. Because the quality of a call is mainly influenced by delay and jitter, the current development implicates new challenges in safety and security which have to be faced.

The term safety describes the ability of a system not to cause environmentally harming events, due to loss of mission critical information – e. g. signaling and media transport in VoIP – under normal and exceptional operations. The term security deals with authentication, authorization, integrity, confidentiality and non-repudiation.

Especially in areas of public safety a packet loss of the voice and signaling stream can result in safety critical events. On the other side the access of unauthorized users who may cause safety issues have to be handled.

This paper gives insight into safety and security in critical communication systems regarding two different abstraction layers - the network and the application layer. As the network layer includes issues typical for IP networks like network failure or security gaps, the application layer has to be investigated towards the use of service

M. Ulieru, P. Palensky, and R. Doursat (Eds.): IT Revolutions 2008, LNICST 11, pp. 186–193, 2009.

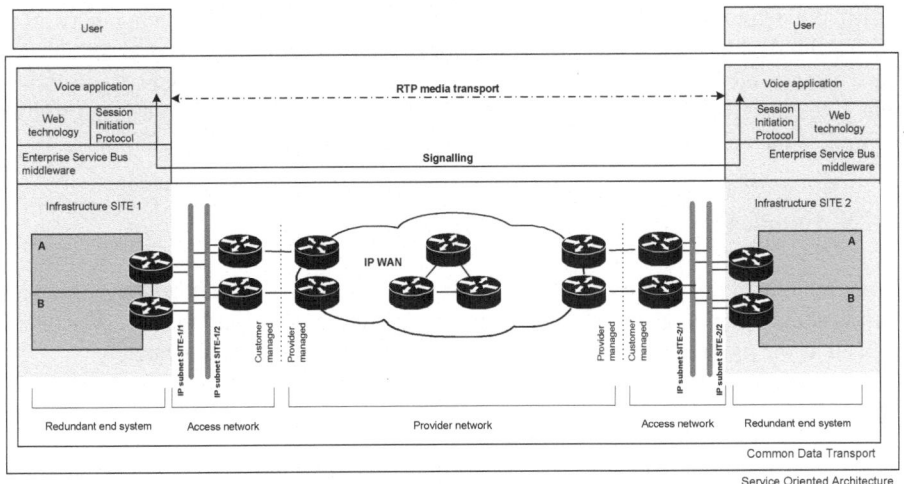

Fig. 1. Service oriented architecture for a safety capable voice service

oriented architecture (SOA). Current developments show a trend towards the use of middleware in order to loosely couple applications and services from their underlying transport system.

Figure 1 sketches different sections of a VoIP communication line which will be in the focus of discussions of this article as well as from the authors' point of view in future evolvements. Figure 1 assumes the use of Internet conferencing architecture protocols – e. g. Real time Transport Protocol (RTP), Session Initiation Protocol (SIP), Session Description Protocol (SDP) standardized by the Internet Engineering Task Force (IETF) – for organizing a VoIP session. Latest developments like the use of SIP by the 3rd Generation Partnership Project (3GPP) for the IP Multimedia Sub-system (IMS) [1] suggest the assumption that SIP increases its impact to the area of VoIP applications.

Contrary to IP, PSTNs already provide a high level of availability to its users. There is no need to develop new standards or mechanisms in order to achieve availability of five nines (99.999% availability) – PSTN already provides this availability for voice communication. RTP, which is used for the payload data transfer in VoIP sessions, bases on the connectionless User Datagram Protocol (UDP) and therefore does not guarantee any reliability or ordering of data packets. Failures on packet transmission can cause considerable interrupts of the voice signal like delayed delivery or loss. IP networks have to be highly available in order to provide safety and need to block any unauthorized access.

Hence, the question has to be asked if the use of VoIP in safety critical systems makes sense if this development is affiliated with a number of problems – the use of VoIP in the area of public safety communication systems seems to be paradox. The reason for its use involves economic considerations as well as service compatibility. Combining different kinds of services on technologically identical networks saves maintenance and operational costs to the operating company respectively the user. In addition, current commercial off -the-shelf (COTS) equipment shows performance to fulfill the requirements for designing highly available and fast reacting networks. Hence the development heads towards critical voice communication over IP and we have to face all related safety and security issues. This paper discusses these issues and gives insight to authors' point of view on this topic regarding the transport layer as well as the application layer.

2 Network Layer

Since the Quality of Service (QoS) of VoIP relies on the underlying network, this section is about upcoming safety and security issues. In the following, the access network as well as the provider network – shown in Figure 1 – are discussed separately.

2.1 Access Network

The access network represents the connection between the end-system and the provider network. As mentioned above Internet conferencing architecture protocols do not ensure reliable data packet delivery, because it would cause transfer delays. Nevertheless, high network availability – five nines regarding to availability and reliability standards of the Eurocontrol [2]   – as well as low convergence time has to be ensured. Even though five nines of availability correspond to a network downtime of about five minutes per year, a sub-second failover time as a maximum has to be achieved in safety critical communication networks, that is, the network can at most fail five minutes per year, but must never fail for more than one second at any given time. Typical IP protocols suffer from poor failure detection times and provide a failover time far above one second due to conservative timing parameters. Technologies like the Resilient Packet Ring [3] on the one hand and proprietary and on the other hand similar protocols like HiPER-Ring [4] or EAPS [5] show convergence behavior far beneath one second. However the requirement of specific hardware or the binding to manufacturers wipe out the advantage of using cheap COTS components and standardized protocols.

In Ševcik et al. [6] different ways to reach low convergence times for the access network represented in Figure 1 have been investigated. The maximum network convergence time is specified with 200 ms. Hardware redundancy is applied to the access network. Using different providers would additionally require proper coordination and an identical IP address plan of both. As Layer-2 protocols are known to show slow network convergence, Layer-3 protocols are investigated for this task within the access network. While protocols like OSPF or IS-IS show convergence times above one second in worst case scenarios it is verified that the combination with the bidirectional forwarding detection (BFD) speed up the network convergence to 160 ms. BFD was launched in 2004 by the IETF for controlling connectivity in the forwarding path and provides failure detection in single-hop, multi-hop, and end-to-end scenarios. In addition, BFD avoids the problem of hiding link breakdowns to the end-system by layer-2 nodes. Sevcik et al. showed that low convergence times can be achieved even thought timing parameters are defined on the side of conservatism due to a software implementation of BFD. The authors' state that the use of BFD allows configuring lower timing parameters and resulting in further decrease of the network convergence duration. Even thought BFD supports short timing parameters, stability issues have to be kept in mind. In [7] it is stated that a reduction may result in higher jitter of the detection time and an increased failure probability by declaring a node as down. BFD is currently released as a draft. However, the authors advance the view that it will become an IETF standard soon. BFD seems to be a step towards low failover times.

2.2 Provider Network

This section discusses upcoming safety and security issues by outsourcing the backbone network to a provider. In Figure 1 the provider-managed network – also called provider network – is symbolized by a cloud. It is characterized by a black box, because the customer has no possibility to modify it. As the customer manages the access network, the SIP registrar, proxy, redirect servers, and the end system, the core network is managed only by the provider. Even though security services like the management of firewalls or virtual private networks (VPN) and the monitoring of QoS can be outsourced to the customer, the provider still has to arrange the routing mechanisms and network security. Sevcik et al. [6] assumed a network failover time of 200 ms for the access network in order to ensure an adequate voice information transfer. The provider network has to match these requirements by guaranteeing a low network failover time using common IP technology.

In circuit switched networks for safety critical communication systems – e. g. air traffic control – routing is generally performed by fixed alternative routing (FAR) algorithms. These algorithms show stability and usability in loosely meshed networks. Failures are handled by redirecting data packets over fixed and predefined direct or alternative routes. Even thought FAR algorithms show some issues in case of multiple simultaneous failures [8] for the use in safety critical communication networks they are preferred to dynamic routing algorithms (DAR). DARs show high complexity and require fully meshed networks to ensure performance. Looking at VoIP, packet switched technologies have different routing requirements. Network protocols generally implement dynamic routing algorithms and therefore implicate avoidable complexity – avoidable regarding safety critical communication networks which generally base on few network nodes compared to commercial provider networks. Multi Protocol Label Switching (MPLS) is a data-carrying mechanism commonly used in core networks. In general it is combined with an Interior Gateway Protocol (IGP) and achieves stability in operation but lacks in network convergence time.

Security is another issue the provider has to deal with when providing a core network for safety critical VoIP communication. Network administrators as well as providers try to protect their networks against unauthorized access and attacks. Distributed denial-of-service (DDoS) attacks represent common threats where the intruder causes a denial of service by attacking a single target within the network. The resulting service downtime has direct influence on safety issues in a safety critical communication system. A possible solution would be to physically separate the safety critical network from the common Internet and therefore eliminate an access from the outside. Hence, the provider has to avoid contact between both networks by directing the traffic over separated infrastructures in the network nodes. This seems to be highly sophisticated and it is questionable if any provider can give a guarantee on a complete physical separation. Additional complexity is introduced by providing a redundant path to the main path like it is done in circuit switched networks to increase network availability. Not only the common IP network has to be separated from the critical VoIP network but also two sub networks have to be introduced. In an ideal case both networks are conducted over different network nodes in order to evade a system breakdown in case of a catastrophic event. One solution would be to use different providers for both sub networks. Beneath additional management of failover procedures and IP address plans it cannot be ensured that different providers will not intersect each other's network at any point.

From the authors' point of view the most likely way to cope with these issues is the construction of an IP network for safety critical VoIP applications without provider infrastructure. Otherwise the listed requirements can hardly be achieved. Even though already available infrastructure cannot be used and upcoming developing costs rise, the advantage of using a uniform and widespread technology, which allows the bonding of numerous services, remains.

3 Service Oriented Architecture in Communication Systems

Trends in telecommunication systems show a convergence of different forms of communication and supporting applications. The idea is building a global Next Generation Network (NGN) based on top of IP [9]. The main challenge of today's companies is to get different platforms together and exchange data when required.

3.1 Overview

SOA [10] defines a type of architecture that abstracts the transport itself to a service definition and can be seen as development of Component Based Architecture (CBA). The service definition targets need to build loosely coupled, standard based, protocol independent and location transparent software components. Services communicate with each other using messages and allow the coordination of activities between them. Business processes are modeled by composition of distributed services shared through a network.

The term service is often confused with Web Service (WS). A WS can be seen as a possible connection technology of SOA. WSs are often used, but do not represent the only possibility for service integration and definition of new services. Services are made accessible by using standard Internet protocols. WSs will play a big role in future VoIP implementations by offering voice communication platforms out of the web browser. Therefore different implementation approaches exist proposing dual (WS using HTTP and SIP) and single stack solutions (purely WS based) [11].

A large part of SOA concepts focus on the implementation of a hub-and-spoke integration pattern realized as an Enterprise Service Bus (ESB) [12]. The ESB represents a Message Oriented Middleware (MOM) allowing simple integration and re-use of business components by the use of open standards. It provides all necessary services like connectivity, message routing, data transformation and adoption to different applications to allow interaction of different distributed resources supporting mediation and orchestration of them. Beside the transport network the software components (e.g. message router, component framework, management) implementing the ESB functionality introduce additional safety critical aspects to the system. Safety has to be considered on distinct levels like message, service or infrastructure. Message transport, service discovery and service interaction are only a few of the many possible safety related issues. It is important that data and events are sent to the right consumer within the required time constraint. Reliable message exchange is needed. It is also of importance that services are available even if components in the network will fail. The ESB infrastructure has to address the problem of isolating faults caused by server and communication infrastructure. Distributed, reliable and available ESB systems need to be developed to allow the usage in safety critical application fields.

Looking at Air Traffic Management (ATM) as a safety critical application area for voice integration, we see that big effort is made to challenge the increasing air traffic capacity demands and enhance safety and security at the same time. The Federal Aviation Administration (FAA) is designing a System Wide Information Management (SWIM) combining data from different sources e.g. flight and flow data, weather information and surveillance using the principles of SOA. In Europe the SESAR project was defined to build the next generation air transportation system where SOA is described as a suitable paradigm to allow Collaborative Decision Making (CDM) [13] by the use of COTS communication and middleware standards. Standards are currently defined to guarantee safe and secure ATM interoperability with VoIP [14]. This paves the path to integrate VoIP services to a SOA based communication network.

This safety critical application field is a good example for lots of other convergence activities, where voice and data converge into a single service capable network. The integration of voice service allows the forwarding of signaling information to a business process management system (BPMS) which handles the call. Service definitions will support presence, call routing, alert generation and data collection. The couplings with other information systems help the operator to make quick and accurate decisions using a clear user interface. Improved workflows and more efficient operations are achievable using intelligent process modeling to build clear situational awareness. Safety objectives identifying minimum requirements to be achieved by the ESB system need to be specified for the different system elements. Consequences by inserting an additional software abstraction layer encapsulating the packet transport have to be investigated carefully to conform to the strict safety requirements.

3.2 Security Issues and Concerns

SOA as an architectural paradigm and discipline for analyzing, designing and implementing distributed systems not only has advantages with respect to efficiency, flexibility and interoperability but also disadvantages in terms of security. Issues and concerns are often disregarded, although they play an important role when talking about safety critical voice and data communication in the field of ATM.

Due to its distributed hardware and software structure, open and manufacturer-specific (proprietary) interfaces, protocols and formats, SOA has new and more comprehensive demands on security with respect to identification, authentication, authorization, confidentiality, integrity and non-repudiation. One by one can be investigated from the perspective of a user and a service. To deal with SOA specific and non-SOA specific requirements, threats and vulnerabilities of valuable assets and liabilities security has to be implied on several levels [15].

In the past, security mechanisms were concentrated on the transport layer. There the security standard Transport Layer Security (TLS) [16] works fine. TLS is wide-spread and its predecessor Secure Socket Layer (SSL) is e. g. used for securing SIP based VoIP communication [17]. However, problems remain with its use in SOA. TLS offers point-to-point security but no end-to-end security. This, however, is the requirement of SOA. The messages are protected only on the transport level. Therefore, security cannot be ensured after the arrival in the end-system. In SOA a message is sent over several intermediary services, which may operate on individual elements of this message. That is the reason why security needs to concentrate on the message level.

The stated weakness of a SOA has to be confronted with new and promising approaches. In other words, a SOA must be supported by a security framework, which reacts to a considerable degree on the varying requirements of the service environment. Hence, a security architecture including identity, authentication and authorization management, intranet, extranet and Internet security, registry security, Universal Description, Discovery and Integration (UDDI) security and messaging security has to be considered [15].

Although a few security requirements can be met with these mechanisms, many are not solved. Though there are many standards e. g. eXtensible Markup Language (XML) [18] and WS security frameworks promoted by World Wide Web Consortium (W3C) [19] and Organization for the Advancement of Structured Information Standards (OASIS) [20], where and how these solutions are applied in a SOA still need further research.

4 Conclusion and Outlook

Reusability and standardization are the direction to which voice communication is heading. We have listed the benefits of this development by using VoIP, but one must not forget the possible fallacies when doing so: when making elementary changes to a technology, the quality of the new system needs to be at least as good as the already available quality. This requires us to think about reliability and availability as well as the voice communication quality itself in terms of delay and distortion. From what we see today VoIP is the way to go and will bring technological benefits as well as cost reduction to providers and customers. But only if we are able to provide at least the same quality as we have today, we will see commercial success and thus replacement of existing technologies. Safety critical areas of voice communication have even higher requirements and in order to meet these requirements, we need detailed analysis of all related technologies.

As a consequence of the convergence of data and voice we are faced with new requirements in the architecture that is capable of enabling this convergence. Moving towards services looks promising and the proposed SOA paradigm is a candidate solution. But again, we must not forget that beside all benefits that we get from this new convergence, the quality that is provided to the user has to remain the same, if not increase.

References

1. Group Services and Systems Aspects: IP Multimedia Subsystem (IMS), Stage 2, 3GPP TS 23.228 Technical specification, Release 7, Version 7.7.0 (2007)
2. European Organization for the Safety of Air Navigation: Voice Communication Procurement Guidelines (2003)
3. IEEE Standard: P802.17-2004 Resilient Packet Rings (2004)
4. Schaub, M., Kell, H.: Produkt-Analyse: HiPER-Ring vs. RSTP Redundanzverfahren mit Hirschmann Switches, ComConsult Research (2003)
5. Shah, S., Yip, M.: Extreme Networks' Ethernet Automatic Protection Switching (EAPS) Version 1. IETF Request for Comments 3619 (2003)
6. Sevcik, B., Zeilinger, H., Zucker, G.: High Available and Reliable IP-Networks for time-critical Voice over IP Applications (to be published)
7. Cisco Systems, Inc.: Cross-Platform Release Notes for Cisco IOS Release 12.0 (28d) (2006)
8. Rausch, T., Zeilinger, H., Kaindl, A.: Routing Performance in Air Traffic Services Networks. In: Proceedings Seventh International Conference on Networking, pp. 669–674 (2008)
9. Blum, N., Magedanz, T.: Requirements and components of a SOA-based NGN service architecture. Elektronik and Informationstechnik, Ausgabe 7–8 (2008)
10. Papazoglou, M.P., van den Heuvel, W.H.: Service oriented architectures: approaches, technologies and research issues. VLDB Journal (2007)
11. Chou, W., Li, L., Liu, F.: Web Services for Communication over IP. IEEE Communications Magazine (March 2008)
12. Ning, F., Xingshe, Z., Kaibo, W., Tao, Z.: Distributed Enterprise Service Bus based on JBI. In: 3rd International Conference on Grid and Pervasive Computing (2008)
13. Eurocontrol: SESAR project, http://www.eurocontrol.be/sesar/
14. EUROCAE: European Organization for Civil Aviation Equipment, http://www.eurocae.eu/
15. Stephens, B.: Security architecture for system-wide information management. In: 24th Digital Avionics Systems Conference (2005)
16. Dierks, T., Allen, C.: RFC2246-The TLS Protocol Version 1.0, Network Working Group (1999)
17. Butcher, D., Xiangyang, L., Jinhua, G.: Security challenges and defense in VOIP Infrastructures. IEEE Transactions on systems, man, and cybernatics – Part C: Applications and reviews 37(6), 1152–1162 (2007)
18. Bray, T., Paoli, J., Sperberg-McQueen, C.M., Maler, E., Yergeau, F.: Extensible Markup Language (XML) 1.0, 4th edn., W3C (2006)
19. W3C: World Wide Web consortium, http://www.w3.org/
20. OASIS: Organization for the Advancement of Structured Information Standards, http://www.oasis-open.org

Paradox in AI – AI 2.0: The Way to Machine Consciousness

Peter Palensky[1], Dietmar Bruckner[2], Anna Tmej[2], and Tobias Deutsch[2]

[1] University of Pretoria, South Africa
palensky@ieee.org
[2] Vienna University of Technology, Austria
{bruckner,tmej,deutsch}@ict.tuwien.ac.at

Abstract. Artificial Intelligence, the big promise of the last millennium, has apparently made its way into our daily lives. Cell phones with speech control, evolutionary computing in data mining or power grids, optimized via neural network, show its applicability in industrial environments. The original expectation of true intelligence and thinking machines lies still ahead of us. Researchers are, however, optimistic as never before. This paper tries to compare the views, challenges and approaches of several disciplines: engineering, psychology, neuroscience, philosophy. It gives a short introduction to Psychoanalysis, discusses the term consciousness, social implications of intelligent machines, related theories, and expectations and shall serve as a starting point for first attempts of combining these diverse thoughts.

Keywords: Machine consciousness, artificial intelligence, psychoanalysis.

1 Introduction

Embedded computer systems have seen their computing power increase dramatically, while at the same time miniaturization and wireless networking allow embedded systems to be installed and powered with a minimum of support infrastructure. The result has been a vision of "ubiquitous computing", where computing capabilities are always available, extremely flexible and support people in their daily lives.

The common view in related communities is that computers will not only become cheaper, smaller and more powerful, but they will disappear and hide or become integrated in normal and everyday objects [21], [22]. Technology will become invisible and embedded into our surroundings. Smart objects communicate, cooperate and virtually amalgamate without explicit user interaction or commands; they form consortia for offering or even fulfilling tasks for a user. They are capable of not only sensing values, but deriving context information about the reasons, intentions, desires and beliefs of the user. This information may be shared over networks – one of which is the world-wide available Internet – and used to compare and classify activities, find connections to other people

M. Ulieru, P. Palensky, and R. Doursat (Eds.): IT Revolutions 2008, LNICST 11, pp. 194–215, 2009.
© ICST Institute for Computer Sciences, Social-Informatics and Telecommunications Engineering 2009

and/or devices, look up semantic databases and much more. The uninterrupted information flow makes the world a global village and allows the user to access his explicitly or implicitly posed queries anywhere anytime.

The vision of machine consciousness – which includes the availability of enough computation resources and devices for fulfilling the tasks – poses many requirements across the whole field of information and communication technology and allied fields [23]. For example, the development in ambient intelligence with respect to conscious environments requires development in the areas of sensors, actuators, power supplies, communications technology, data encryption and protection, privacy protection, data mining, artificial intelligence, probabilistic pattern recognition, chip design and many others not stated here. Each research group and even researcher has its own view on what machine consciousness will be or will approximate. The same is already true for the many necessary technologies required for machine consciousness. The research on ambient intelligence for instance can be divided up [24] into three basic research areas – or existing research projects can be grouped into projects investigating three basic methods – which reflect fundamentally different approaches in establishing ubiquitous computing environments: Augmented Reality (see for example [25]), Intelligent Environments (e.g. [26], [27], [28]), and Distributed Mobile Systems (see [29], [30]).

The border between smart devices and devices with machine consciousness may be something very controversial and from the philosophic point of view of massive importance and implications. This will be true for ambient intelligence applications as well as any other, e.g. industrial, conscious application. However, for the user of such a device this distinction is meaningless: A user expects the fulfillment of a particular task, the device just needs to be intelligent enough to to so. An argument for calming down the apprehension of machines that will control us is presented below with the model of nested feedback loops, each loop representing a bit more consciousness than the lower ones. In this context the designer of a machine can control the level of consciousness he wants from his device.

These considerations together give a rough definition of machine consciousness as we see it: Machines equipped with decision units that allow them to think in a way humans do. The key issue is the "thinking in a way humans do" in opposition to "acting looking like humans" of many well-know projects, which is nothing but mimicry. There are, e.g., robots capable of performing facial expression like smiling. However, this does not imply that the robot "is amused" as a human would be who smiles, or that the robot has any other human-like intentions to smile. But this would be the necessary requirement for attributing this robot consciousness.

2   What Consciousness?

Consciousness[1] helps us humans to put ourselves into the set of parameters and actors when decisions are made. A concept of self, its desires and plans is one attribute that makes someone appear intelligent.

---

[1] The interested reader might be referred to David Chalmers' and David Bourget's web repository http://consc.net/online, which lists more than five thousand papers on consciousness.

Most of all, consciousness is a subjective quality. Person A can feel its own consciousness, be aware of it, test it and have a concept of itself. Person A can tell Person B about this extraordinary experience but Person B can never be sure. Only from "inside", consciousness can be experienced and verified. The link of this qualia to the physical process, the physical machine, is unfortunately not clear [35].

The outer shell of a conscious being might exhibit very distinctive behavioral patterns. These patterns might in turn be checked against a "turing test on consciousness". There are, however, numerous arguments about machines, potentially passing these tests but still being "zombies", i.e. lacking consciousness [36].

Another distinction is where we assume consciousness. Generally accepted as a human phenomenon (maybe also in "higher" mammals), its projection to or implementation in machines [37] opens up two problems:

1) We could be fooled by anthropomorphic aesthetics. Humans actively seek emotions in faces, even if they are inanimate rubber masks and we should be aware of this pitfall.

2) If we created this machine, we have total power over its hardware (and software, if we use contemporary terms) and can copy it, switch it off and on without damage, modify it, monitor it without interference, etc.

It is the second point that would make these beings massively different to humans. We have no power over our own hardware. We are mortal and do not fully understand how our body works. Out of this situation we can derive two alternative outcomes:

a) A machine that can be switched off and on cannot host consciousness because it would be far too primitive. Once we have machine consciousness we will realize that the hardware has been gradually taken out of our hands (e.g. manufactured by nano-bots and evolutionarily modified beyond our knowledge) and therefore as intangible as our own brains.

or

b) It works and constitutes one half of what many people dream of: a potential "storage" or platform for our own consciousness, ultimately leading to immortality (the other half, however, is still missing: understanding of our own hardware a.k.a. "wetware", to download its content to the new platform). Unfortunately, a subjective engineering estimate would be that the chances that such a machine – if it develops something like consciousness – is compatible to our "software" are virtually zero.

So we might reach some point where we have remarkable and potentially conscious artifacts, but their comparison to our own body and mind as a unity is very questionable, although most technical applications would be happy with the mind of a sophisticated "zombie" [42].

3 Consciousness Support

To provide a machine with machine consciousness one must, in order to stay scientifically consistent, review already existing research on human consciousness.

In this chapter we will give an overview about how various sciences approach consciousness. For technical purposes a holistic, functional model without contradictions is desired. We will choose an existing method as a template and go into more depth into that theory.

3.1   Philosophy

The questions of consciousness that philosophy traditionally dealt with can be gathered in three crude rubics as the What, How and Why questions: What is consciousness, its principal features? How does consciousness come to exist, and finally, Why does consciousness exist? David Chalmers [7] summarizes more modern philosophical approaches when he differentiates between the so-called "hard problem" and the "easy problem" of consciousness. The latter concerns objective mechanisms of the cognitive systems: the discrimination of and reaction to sensory stimuli, the integration of information from different sources and the use of this information to control behavior and the verbalization of internal states, while it is the "hard problem" that actually deals with the "mystery" of consciousness [7], p. 62: the question of how physical processes in the brain give rise to subjective experience. This involves the inner aspect of thought and perception: the way things feel for the subject. This part of consciousness is also called "phenomenal consciousness" and "qualia" [8], while Chalmers' "easy problem" is also called "access-consciousness". Daniel Dennett [9], on the other hand, denies that there is a "hard problem", asserting that the totality of consciousness can be understood in terms of impact through behavior. He coined the term "heterophenomenology" to describe an explicitly third-person scientific approach to human consciousness.

Dennett's new term is closely linked to his cognitive model of consciousness, the Multiple Drafts Model (MDM) [9]. According to this model, there are a variety of sensory inputs from a given event and also a variety of interpretations of these inputs. The sensory inputs arrive in the brain and are interpreted at different times, so a given event can give rise to a sequence of discrimination, constituting the equivalent of multiple drafts of a story. As soon as each discrimination is accomplished, it becomes available for eliciting a behavior. Like a number of other theories, the Multiple Drafts Model understands conscious experience as taking time to occur. The distinction is that Dennett's theory denies any clear and unambiguous boundary separating conscious experiences from all other processing. According to Dennett, consciousness is to be found in the actions and flows of information from place to place, rather than some singular view containing our experience. The conscious self is taken to exist as an abstraction visible at the level of the intentional stance, akin to a body of mass having a center of gravity. Similarly, Dennett refers to the self as the center of narrative gravity, a story we tell ourselves about our experiences. Consciousness exists, but not independently of behavior and behavioral disposition, which can be studied through heterophenomenology.

3.2 Psychology

Before the advent of cognitive psychology, psychology failed to satisfactorily study consciousness. Introspectionism, first used by Wilhelm Wundt as a way to dissect the mind into its basic elements, dealt with consciousness by self-observation of conscious inner thoughts, desires and sensations. The relation of consciousness to the brain remained very much a mystery. This experimental method was criticized by upcoming behaviorists as being unreliable; scientific psychology should only deal with operationalizable, objectifiable and measurable contents. Behaviorism thus studied the mind from a black-box-point-of-view, stating that the mind could only be fully understood once the inputs and the outputs were well defined, without even hoping to fully understand the underlying structure, mechanisms, and dynamics of the mind.

Since the 1960s, cognitive psychology has begun to examine the relationship between consciousness and the brain or nervous system. The question cognitive psychology examines is the mutual interaction between consciousness and brain states or neural processes. However, despite the renewed emphasis on explaining cognitive capacities such as memory, perception and language comprehension with an emphasis on information processing and the modeling of internal mental processes, consciousness remained a largely neglected topic until the 1980s and 90s.

A major example of a modern cognitive approach to consciousness research is the global workspace theory of Bernard Baars [1]. It offers a largely functional model of consciousness which deals most directly with the access notion of consciousness and has much in common with the multiple drafts model mentioned above. The main idea of global workspace theories is that consciousness is a limited resource capacity or module that enables information to be "broadcast" widely throughout the system and allows for more flexible sophisticated processing. It is thus closely allied with many models in cognitive psychology concerned with attention and working memory.

Just as the philosophical approaches mentioned above, however, psychological attempts at describing and investigating consciousness lack a thorough and exact analysis of the structure, mechanisms and dynamics of consciousness and mental processes.

3.3 Evolutionary Biology

Evolutionary Biology has mainly investigated the question of the causes for consciousness. From this point of view, consciousness is viewed as an adaption because it is a trait that increases fitness.

3.4 Physics

Modern physical theories of consciousness can be divided into three types: theories to explain behavior and access consciousness, theories to explain phenomenal consciousness and theories to explain the quantum mechanical (QM) Quantum mind [43], [44]. These latter theories are based on the premise that quantum mechanics is necessary to fully understand the mind and brain to explain consciousness. The quantum mind hypothesis proposes that classical mechanics cannot fully explain consciousness and suggests that quantum mechanical phenomena such as quantum entanglement and superposition may play an important part in the brain's function and could form the basis of an explanation of consciousness.

3.5 Cognitive Neuroscience

This scientific branch is the most modern approach to consciousness research, primarily concerned with the scientific study of biological substrates underlying cognition, with a specific focus on the neural substrates of mental processes and their behavioral manifestations. Amongst other things, it investigates the question of how mental processes uniquely associated with consciousness can be identified. It is based on psychological statistical studies and case studies of consciousness states and the deficits caused by lesions, stroke, injury, or surgery that disrupt the normal functioning of human senses and cognition. Cognitive neuroscience is a branch of both psychology and neuroscience, unifying and overlapping with several sub-disciplines such as cognitive psychology, psychobiology and neurobiology. One major question that cognitive neuroscience deals with is the so-called "mind-body-problem" [17], the question how brain and mind or brain and consciousness relate to each other. Many cognitive scientists today hold the view that the mind is an emergent property of the brain: mind and brain equally exist, however they exist at different levels of complexity.

Antonio Damasio [2] differentiates between "core consciousness" and "extended consciousness". While core consciousness describes a hypothesized level of awareness facilitated by neural structures of most animals, which allows them to be aware of and react to their environment, extended consciousness is a much more complex form of consciousness, allowing for a sense of identity and personality and meta-consciousness, and linking past, present and future. Higher forms of extended consciousness only exist with humans and depend on (working) memory, thinking and language. Just as all other scientific approaches mentioned above, cognitive neuroscience, too, does not provide us with a detailed and concrete model of the mental processes involved in the structure and mechanisms of consciousness.

3.6 Psychoanalysis

Psychoanalysis opened up the research field of consciousness to the unconscious properties of the mind. As described in more detail below, Freud developed his metapsychological ideas from a theory of three mental processes – conscious, preconscious and unconscious – to a theory of three agencies – the Ego, the Id and the Super-Ego – functioning on the basis of these processes. Although Freud developed his theory as a contrast to the psychology of consciousness exclusively current at his time [10], naturally, the mind that Freud set out to analyze and describe with great precision includes and determines consciousness: although many thoughts or other psychic/mental contents may never reach consciousness, they will always exert influence on consciousness and behavior. According to Freud, it is therefore both legitimate and necessary to include those properties that lie behind consciousness within our conception of the mind [17].

Freud and other psychoanalysts after him saw and see consciousness largely as a means to perceive outer and especially also inner events and inner-psychic qualities [31], as a property of the mind as opposed to the mind itself [17]. The mind, the mental apparatus, including conscious and unconscious properties was described by Freud in all its functions and dynamics. In the end, psychoanalysis emerges as the only science dealing with human consciousness on a detailed enough level to be used for an implementation of machine consciousness.

4 Psychoanalysis, the Template

For the scope of this work psychoanalysis was chosen as the template theory of human consciousness. The main reasons therefore are the functional approach of modelling human thinking, which very well fits a computer engineering approach, and the somewhat technical or natural scientific approach of Freud to the topic which resulted in texts and arguments that can be followed and agreed on by natural sceintists. This chapter gives an introduction to people not from the field.

4.1   Basics

Psychoanalysis was founded by Sigmund Freud (1856-1939), originally a neurol-ogist and neuroanatomist [6], who in his study of hysteria developed his first ideas about unconscious affects, psychic energy and the cathartic effect of verbal expression. In the course of his life, Freud developed these and other concepts further, abolished some, modified others, while inspiring many other scientists to join him in his quest. Some of these came to disagree with Freud in time and went on to pursue their own strands of theory.

Freud himself [12] described psychoanalysis as follows: "Psycho-analysis is the name (1) of a procedure for the investigation of mental processes which are almost inaccessible in any other way, (2) of a method (based upon that investigation) for the treatment of neurotic disorders and (3) of a collection of psychological information obtained along those lines, which is gradually being accumulated into a new scientific discipline" (p. 235). He goes on to differentiate: "The assumption that there are unconscious mental processes, the recognition of the theory of resistance and repression, the appreciation of the importance of sexuality and the Oedipus complex – these constitute the principal subject-matter of psycho-analysis and the foundations of its theory" (p. 250).

After Freud died in 1939, psychoanalytic theory and practice continued to be further developed by scientists and practitioners as Heinz Hartmann, Anna Freud (Sigmund Freud's youngest daughter), and Melanie Klein, to name just very few. However, already before Freud's death, there had been disagreements about the meaning of certain concepts or ways of treatment. Since then, contro-versy has continued to be characteristic of psychoanalytic theory; distinct schools of thought focusing on different topics, e.g. Ego-psychology or object relations have emerged. Even today, more than 100 years after its foundation, psychoanalysis is a living science in the sense that the different concepts of the different schools are still being further developed and discussed. While psychoanalysis is for a great part concerned with psychotherapeutic methods, psychopathologies and individual developments leading thereto, one great part also consists of metapsychology, or conceptualizing the psychic apparatus and the way it (mal-) functions. As this part of the psychoanalytic theory shall be our main focus, this introduction will simply exclude the other, albeit certainly very important and also characteristic aspects of psychoanalysis.

Additionally, although, as described above, there are many different strands of theory in psychoanalysis, each with different main focuses and different understandings of certain concepts, in this introduction, we will still concentrate on Freud's original conception of psychoanalytic theory. This is due to the very fundamental and basal nature of Freud's original ideas which to this day have not been abolished, and continue to be valid also in modern psychoanalytic theory, although different schools may focus on different concepts from different periods in Freud's life and development of the theory. Details about the technical conceptualization and partly implementation thereof can be found in [40]. The authors there in a first attempt concentrated on employing the original psychoanalytic theory, before venturing further to include other, more modern psychoanalytic concepts, be it the case that these prove as usable and expedient as Freud's conception of the psychic apparatus.

4.2   The Mental Apparatus (1): The Topographical Model

Freud conceptualized the mind as being divided up into different parts, each the home of specific psychological functions. His first model, the topographical model [14], divides the mind into the unconscious, the preconscious and the conscious system. Characteristic of these different parts are two different principles of mental functioning [11] – the so-called primary and secondary processes. While secondary process thinking – typical of conscious processes – is rational and follows the ordinary laws of logic, time and space, primary process thinking – typically unconscious – is characteristic of dreaming, fantasy, and infantile life in which the laws of time and space and the distinction between opposites do not apply. Some psychological processes are however only unconscious in the descriptive sense, meaning that the individual is not aware of them, but they are easily brought to mind (preconscious). With dynamically unconscious processes, on the other hand, it is not by simple means of effort or change of attention that they can be rendered conscious. These to the conscious system unacceptable psychic contents are subject to repression and operate under the sway of primary processes.

4.3   Drive Theory

Freud saw the internal world as dominated by man's struggle with his instincts or drives. In his initial formulation of instincts, Freud [15] distinguished between self-preservative (e.g. hunger) and sexual drives (libido). Later, he stressed the difference between sexual drives and aggressive or destructive drives (Eros vs. Thanatos) [16]. Classically, instinctual wishes have a source, an aim, and an object. Usually, the source of the drive is infantile and lies in the body, possibly in an erogenous zone. Over time, after several similar (real or imagined) satisfactions of instinctual wishes, source, aim and object begin to mesh together into a complex interactional fantasy, part of which is represented in the system unconscious.

4.4   The Mental Apparatus (2): The Structural Model

In the structural model, Freud [12] proposed three parts or structural components of the human mind: Id, Ego and Super-Ego.

Id

The Id is the first psychic structure of the mental apparatus, out of which, in the course of infantile and childhood development, the Ego and the Super-Ego evolve. It contains the basic inborn drives and sexual and aggressive impulses, or their representatives. As such, it is an inexhaustible source of psychic energy for the psychic apparatus: The Id's wishes strive for immediate satisfaction (pleasure principle) and therefore drive the functions of the Ego to act. Its contents are unconscious and function following the primary process.

Ego

In the course of infantile development, the perceptive and executive parts of the Id, responsible for drive satisfaction by perceiving the baby's environment and ways to gain satisfaction, start to form a new part of the mental apparatus: the Ego. The mature Ego's tasks, however, exceed perception and execution: the Ego has to control the primitive impulses of the Id and to adapt these to outer reality (reality principle) as well as to mollify the requirements of the Super-Ego. For these purposes, the Ego makes use of so-called defense mechanisms such as repression to keep unacceptable impulses within the Id and therefore evade conflict with either outer reality or Super-Ego-requirements. The contents of the Ego are partly conscious, partly unconscious. A however certainly not complete list of Ego functions should not omit the following [5]: consciousness; sensory perception; perception and expression of psychic agitation; thinking; controlling motor functions; memory; speech; defense mechanisms and defense in general; fighting, controlling and binding drive energy; integrating and harmonizing and reality check.

Super-Ego

The Super-Ego comprises the conscience and ideals, thus allocating (moral) rules and prohibitions which are derived through internalization of parental or other authority figures, and cultural influences from childhood onwards. The Super-Ego resembles the Ego in that some of its elements are easily accessible for consciousness, while others are not. Super-Ego ideation also comprises rational and mature presentations up to very primitive and infantile ones. The task of the Super-Ego is to impact on the actions of the Ego, especially to support it in its defensive actions against the drives with its own moral rules. However, the relationship between the Ego and the Super-Ego will not always be this harmonic: in other cases, e.g. if the difference between instinctual or other repressed wishes from the Id and moral rules from the Super-Ego becomes to great, the Super-Ego produces feelings of guilt or a need for punishment inside the Ego.

4.5 Psychoanalysis and Neuroscience

In recent years, a new scientific research strand has developed aiming at supporting and reassigning psychoanalytic concepts by neuroscientific findings: neuropsychoanalysis. Solms [18] provides a summary of the neuroscientific support for some basic psychoanalytic concepts, one of which is the notion that most mental processes occur unconsciously. Different memory systems have been identified, some of which are unconscious, which mediate emotional learning. The hippocampuses, which lay down memories that are consciously accessible, are not involved in such processes. Therefore, no conscious memories are available and current events can "only" trigger remembrances of emotionally important memories. This causes conscious feelings, while the memory of the past event remains unconscious. In other words, one consciously experiences feelings but without conscious access to the event that triggers these feelings.

The major brain structures for forming conscious memories do not function within the first two years of life. Developmental neurobiologists largely agree that early experiences, especially between infant and mother, fundamentally shape our future personality and mental health. Yet none of these experiences can be consciously remembered. It becomes increasingly clear, that a great deal of our mental activity is unconsciously motivated.

Freud's ideas regarding dreams – being instigated by the drives and expressing unconscious wishes – were at first discredited when rapid-eye-movement sleep and its strong correlations with dreaming were discovered. REM sleep occurred automatically and was driven by acetylcholine produced in a "mindless" part of the brain stem, which has nothing to do with emotion or motivation. Dreams now were regarded as meaningless, simple stories concocted by the brain under the influence of random activity caused by the brainstem. Yet recent work has revealed that dreaming and REM sleep are dissociable while dreaming seems to be generated by a network of structures centered in the forebrains instinctual-motivational circuits. These more recent views are strongly reminiscent of Freud's dream theory.

These developments and the advent of methods and technologies (e.g.: neuroimaging) unimaginable 100 years ago make it possible today to correlate the psychoanalytic concepts, derived from observation and interpretation of subjective experiences, with the observations and interpretations of objective aspects of the mind studied by the classical neurosciences, thus rendering psychoanalysis and psychoanalytic theories objectifiable.

5   Road-Map

The capabilities of the desired machine consciousness introduced above represent a development stage in the future, which requires several disruptive innovations. We cannot say today how long it will take to reach them. There are many other authors predicting human-like intelligence within some decades, or only in 100 years, or even never. The authors here believe we are much closer! We see many promising approaches to problems of AI, their only drawback lies in their position within the framework of the application – which, in our opinion, shall be given by psychoanalysis [40]. However, psychoanalysis is not concerned with all the necessary functions, it is just and primarily concerned with psychic functioning itself.

Psychoanalysis requires so-called memory traces. A memory trace is the mental representation of something, be it a person, an object, event or whatever. Psychoanalysis is not concerned with perception (which generates memory traces), or with direct motor control. For technical systems these are however key requirements, needless to say these are the fields where robotics, AI (e.g. Software Agents) and others put most of their efforts in, but we need to search for other templates than psychoanalysis. Candidates here are developmental psychology and neuroscience, to mention just two. It becomes more and more clear that a system showing something like machine consciousness needs to be designed following a model, of which the mentioned disciplines have different viewpoints. Therefore, it needs to combine them in a holistic manner.

Psychoanalysis together with the other mentioned fields span a huge search space for engineers. Therefore it is not possible (and useful) to start trying to implement all of it. It is necessary to define a road map with milestones and priorities. A good hint is given by Rodney Brooks in his recent Spectrum article [34], where he introduced four capabilities of children, which are vital behavioristic observable capabilities of human-like intelligent machines. Unfortunately, these have nothing to do with psychic functioning; they are just observations of behavior. There may however be one connection in the last point – the "theory of mind" which in psychoanalysis is called "mentalization" or "reflective functioning" [39], which is definitely a crucial capacity within AI (being able to guess what another person is thinking, planning; understanding why a person does whatever it is they are doing etc.).

5.1   Mental Functions

Before mentioning and commenting Brooks' list, it is necessary to explain some of the very basic concepts of psychoanalysis that need to be addressed by the first machines with comparable behavior. These are the primary and secondary process, terms that refer to the quality of thoughts – or mental content in terms of uncontrolled, impulsive actions or considered ones. The functions necessary for this according to [19] are:

perception: The mind knows two kinds of perception, internal and external. Internal perception results from an observational perspective on the mind. Both together allow for mental content of the kind: "I am experiencing this".

memory: The mind recognizes previous mental experiences of the mentioned kind. It is also able to derive cause-and-effect sequences thereof. These two capabilities together form the immature ego, which only depends on drives and environment.

emotions: The very basic function of emotion is to rate something in terms of good or bad in a biological sense. More biologically successful actions are felt satisfying. Emotions are used to rate the above mentioned perceptions and memory content. In this way, quantitative events acquire quality.

feelings: Feelings are the base for consciousness. Their function is to evaluate the other mechanisms. The kind of mental content that results is "I feel like this about that". Feelings are also stored together with their generating mental experiences. They can be used to create motivation by trying to repeat previous experiences of satisfaction. Motivation searches through past experiences and matches them with the present to come to decisions about what is to be done. The decisions lead to actions (motor output). These all together form the primary process.

inhibition: Experience about actions show that some of them are only satisfying in the short-term, but not in the long-term. Therefore it is necessary to be able to tolerate temporary unpleasure via inhibiting the immediate action plan. This capacity permits thinking. Thinking is inhibited (imaginary) action, which permits evaluation of potential (imaginary) output, notwithstanding current (actual) feelings. This function is called secondary process. It replaces immediate actions with considered ones.

The ability of the secondary process develops very early in a child at the age of around two. However, the finesse and long-term anticipation capabilities in imaginary actions develops during the whole life.

The above description lets us conclude that the mind is organized as a system of nested feedback loops. If we leave memory aside, the first level consists of perception and evaluation. Let's call the pure perception mental content level 1^2. The evaluation together with the perception, towards it is targeted, are again stored. This mental content is another piece of information; let's call it mental content level 2. Content of level 2 is rated with feelings. These together form mental content level 3. The concept of evaluating mental content with feelings from level 3 upwards is the same for the all upper layers. It is not clear at this point if there are intermediate levels, but one of the next levels is related to language. With this, some mental content reaches a level of attention which is important enough to give it a name or a phrase – and therefore being able to share it with others. Again, in the upper levels, reasoning and feeling about mental content related with language – dialogs, names, etc. – form the next levels.

It is important that in this concept the lower levels of some mental content need to be "created" before upper levels can even be thought of. This may be

---

[2] Here we need to mention that "perception" itself requires lots of computation before generating the above mentioned memory traces. So, in a holistic model, we would start with level 1 on the sensor level.

one reason why it is so hard to bring findings or knowledge of different context together.

5.2 Example

As an example let us think about the following situation: Somebody (being it a child, an ancient hunter, whoever) is walking in nature and sees a steep hillside, an edge. Mental content level 1 would be "I see an edge". For some reason (maybe searching for food, playing, etc.) this edge generates the interest of the person, letting it think (conscious or not) "I am interested in that edge". Therefore, the person moves itself to that place. This involves basic action planning. However, when the person reaches the edge, the region behind gets visible. Let us further assume the person sees something he has searched for. Then mental content of the kind "I like this place – and also the landmark (the edge) which indicates the place" could arise, because walking there results in reaching something desired. If this desired thing is interesting or important enough, the person could think "I would like to tell my spouse of this" and "I call this region (e.g.) the XYZ place and the landmark the RST edge".

In this way, perception is enhanced with evaluation level by level. The more interest (positive or negative) it creates, the more chances for further processing are implied. Things which cannot be perceived (because of a lacking template) or which generate neutral evaluation are considered unimportant.

In the following the four capabilities mentioned by Brooks are described (the first statements of the following sections marked with quotes). We try to formulate necessary psychic functions related to the introduced levels of mental content that enable the observation of the desired behavior. Additionally, other necessary concepts for rudimentary machine consciousness beyond Brooks' ones are introduced.

5.3 The Object-Recognition Capabilities of a 2–Year–Old Child

"A 2–year–old can observe a variety of objects of some type – different kinds of shoes, say – andsuccessfully categorize them as shoes, even if he or she has never seen soccer cleats or suede oxfords. Today's best computer vision systems still make mistakes – both false positives and false negatives – that no child makes."

Children recognize shoes, not their size, weight, etc. They recognize shoes because of their roughly similar shape and their functionality. That means: children know what shoes in general look like (foot-shaped plus a place where the foot can be inserted) and what they are for (that is: for putting them on and walking around). The rest is variable: exact shape, color, price; and does not interfere with basic shoe-recognition. Therefore, also a giant, two-meter shoe placed on the top of the roof of a giant shoe-shop will be recognized as a shoe, although it technically is, of course, not a shoe (but: it is foot-shaped, even though giant, and you could theoretically put your foot into it and therefore also walk around in it); however the child is naive enough not to know about other things that would hinder this shoe from really functioning in a shoe-like manner.

In visual object recognition requirements this would translate into many prob-: lems, an algorithm would need to solve: Finding out the geometry of the picture. How large is the area covered, how large are the objects. There needs also to be a model of the object in 3D. The visual classificator has to be based not only on features representing the spatial change in e.g. luminosity of pattern, but needs to take contours into consideration. A promising approach in this respect could be a combination of the boosted cascade of feature detection [20] relying on a set of features like shape, color, etc., not only pattern, as described e.g. by [41].

The very important aspect however is deriving the meaning of things for the machine. What can the machine do with that object? Can it be evaluated good or bad (or some more categories) in terms of the application of the machine?

5.4   The Language Capabilities of a 4–Year–Old Child

"By age 4, children can engage in a dialog using complete clauses and can handle irregularities, idiomatic expressions, a vast array of accents, noisy environments, incomplete utterances, and interjections, and they can even correct nonnative speakers, inferring what was really meant in an ungrammatical utterance and reformatting it. Most of these capabilities are still hard or impossible for computers. "

As elaborated above, the meaning is the essential thing. Language is a data transmission channel to allow the creation of mental content in the mind of the listener as desired by the speaker. It is clear that the rules of grammar and language (word types, irregularities, categories, etc.) need to be learned (stored), however, language represents another level of quality for mental experiences in terms of perception, memory, emotions, and feelings. A lot of learning and personally evaluating of lower level mental content is involved before.

5.5   The Manual Dexterity of a 6–Year–Old Child

"At 6 years old, children can grasp objects they have not seen before; manipulate flexible objects in tasks like tying shoelaces; pick up flat, thin objects like playing cards or pieces of paper from a tabletop; and manipulate unknown objects in their pockets or in a bag into which they can't see. Today's robots can at most do any one of these things for some very particular object. "

The addressed capabilities are closely related to imagination. Imagination is the reasoning about imaginary actions. It requires a large storage pool of already perceived actions and thoughts together with their evaluation in various respects. On the other hand, those capabilities require perfect motor control – which is something where machines can finally (in the context of this paper) compete with or even outperform humans.

5.6   The Social Understanding of an 8–Year–Old Child

"By the age of 8, a child can understand the difference between what he or she knows about a situation and what another person could have observed and therefore could know. The child has what is called a "theory of the mind" of the other person. For example, suppose a child sees her mother placing a chocolate bar inside a drawer. The mother walks away, and the child's brother comes and takes the chocolate. The child knows that in her mother's mind the chocolate is still in the drawer. This ability requires a level of perception across many domains that no AI system has at the moment."

Actually, this is the only point really in connection with psychoanalysis – as a theory about the adult mind. (see above), and being able to "mentalize" or having a "theory of mind" really is a crucially important capacity for any social interaction to succeed: we always consider the other's point of view: what they might be thinking or feeling at the moment (always being aware of the fact that we will never be able to fully know what it is they are thinking or feeling), which mental states make them behave in a certain way, or considering that they do not know what we know. This capacity works in the background and is certainly not consciously intended, but still it is there and absolutely necessary for our understanding of ourselves, other persons and the interactions we come upon.

6   Necessary Basic Concepts

The above depicted road-map takes some concepts as granted. Among them are embodiment, a complex environment and a social system. This section gives a short overview on these topics.

Following the concept of "embodied cognitive science" as described in [32], intelligence must have a body that again shapes the way an agent is able to think. The body can be clearly distinguished between the surroundings, although the body has to contain an interface that grants bidirectional communication to the surroundings. Therefore, the body contains different sensors to sense the surroundings of the agent. Compared to the human's body, the sensors take over the functionality of our five senses (taste, sound, tactile, sense of smell and vision) but have, due to the duty of the agent other functionality. To directly interact with the environment and fulfill the requirement of proactiveness of an autonomous agent, it also has to be equipped with actuators. The more complex the actuators are, the higher is the degree of freedom, the agent can interact and the more possibilities exist to reach desired goals. To lend importance to the body, it is equipped with internal sensors, that are monitoring internal values that are, together responsible for the homeostasis of the agent. This can be in terms of robotic agents e.g. energy level, processor load or internal network load, responsible for fast and slow internal message systems, comparable to the humans hormone system. Since, an agent shall be able to learn, also direct feedback through the internal sensors is eligible as a result of taken actions. Therefore, the environment has to contain dangerous components to gain deeper insight into its own body and allows approximating limits that are given in the environment. Each agent is placed into an ecological niche – defined by the environment and the possibilities of the agent to interact with it. According to Brooks [33], intelligence is within the environment and not within a problem solver. Thereafter, an agent is as intelligent as manifolded and/or complex its senses and actuators are.

An environment in which agents operate has to contain enough distinguishable objects and areas to enable the agent to navigate within it. Like watchdogs have good understanding where home and where outside is, software agents could then develop similar abilities. Also emotional cathexis of objects leads in a complex world to better results. If only few different obstacles are available, almost every place will have the same carthexis. The home will then be very positive, areas close to enemies are emotionally difficult.

The social understanding of an 8 year old as depicted above is only a small fraction of the complex social structures present in the human world. With added complexity and intelligence of (multi-) agent systems dynamic societies are needed. They are less complex as human ones, but nevertheless comparable to them. A society needs social rules (predefined or emergent). Social rules can be help others, don't go there, etc. they glue the society together. They are also some kind of reward system – he who owns great social reputation may receive more help and has more possibilities to influence others. Different societies have different characteristics: some value the experience of the elders, some the creative approaches, some strong leaders, etc. This is dependent on the needs of the society. A further advantage of societies and their social rules and social system is the possibility to specialize. One agent on its own has to do everything, in combination with others he can specialize into one task. This could also be done by design of a special task agent, but the specialization of an all purpose agent into a few tasks has the advantage that if other agents fail to deliver the required product/task (which has been agreed upon via the social system) the agent can do it by himself – albeit in lower quality or slower.

7 Implementation Issues

Implementing traditional AI demands for mathematical skills, using optimized algorithms and data structures. The above concepts, however, are very heterogeneous and have a rich structure. We have no reason to believe that an implicit architecture might be trained to show the desired behavior. Therefore the possibility of an explicit implementation shall be assumed here. Explicit in this context means that the various functional components and mechanisms, as described above, find their direct manifestation in a technical artifact – let's assume functional code.

Theoretically any technical function, described by a formal language, can be implemented solely in hardware – or entirely in software. This assumption holds, as long as we move withing the von-Neumann world of machines. Taken as a black box, no-one cares about the engineering principles inside, as long as it behaves on its interfaces as specified. Generating a specific mathematical function might be done via a lookup-table or via a formula, while both can be implemented and executed both in hardware or software (executed on an abstract, universal machine). The level of abstraction that the implementation shows towards the real meaning is sometimes called the "semantic gap", discussed in [4]. Its consequence is decreased performance. The larger the gap, the worse the performance. This derated performance can take extreme dimensions when the problem contains chaotic aspects. The human mind in general show extremely complex, nonlinear and chaotic behavior, stemming from incredibly large amounts of weighted memories that take influence on decisions. This exposed chaos gives an impression on the complexity of the underlying information processing. An implementation will face the same complexity and therefore strongly benefit from an as small as possible semantic gap. The following "software-near" hardware are examples for typical building blocks used in "small-gap machines":

- Associative memory with database-like capabilities in hardware
- HW support for matrix operations
- Dynamically adaptive microcode/hardware
- Non-blocking multi-port memory
- Merging of memory and operations, hardware-in-memory
- Multidimensional, and not linear, connection of data and functions

All this is of course a violation of the pure von-Neumann idea of an abstract machine that outsources all complexity and functionality into time and software code. As current computers are (by implementing aspects of the above list) already far away from this pure idea, this philosophical aspect should not bother us.

In general, the above ideas of an intelligent machine demand a high level of parallelism in software and subsequently in hardware, if we want operations optimized. Parallelism, in turn, is a challenge to our technology, captivated in the boundaries of space and time. 3D-silicon might be one step more but the level of parallelism we need for instantaneously and successfully querying a tremendously large memory reaches beyond that.

Generally we can expect two sources of innovation that might bring us to the stage to implement our abstract concepts

- Technological innovation
- Structural innovation

Even nowadays and throughout history these two phenomena have – typically in alternating roles – brought technology forward. Just think of the development of computers where technological progress like relais, tubes, TTL, CMOS etc. were – usually when the current technology was on its limits – interlaced with structural innovations like parallelism, superscalar architectures or multi-cores. We can imagine massively parallel description languages that manifest in molecular nanotech and other currently unbelievable technologies to be the platform of tomorrow. Structuring the functional blocks of our models, however, will still be necessary. Taking X Billions of neurons does not create a human brain. It is the structural order which makes it special.

One open question is the scalability of this concept. Implicit AI typically has good scalability characteristics. A distributed genetic algorithm might for instance be scaled via its population and genome size. Similar things can be said of artificial neural networks, but as discussed in [3], neural networks do unfortunately not surprise us with unexpected smartness if we allow for growing.

The capabilities of ANNs do not really aggregate. If "more" is wanted, the ANN must show some macroscopical structure. The above concept's structure is, however, macroscopically not scaling. It does not necessarily get twice as smart just if you introduce two semantic memories than just one. The size of the individual components, especially filter networks or memories, surely can grow if the given problem demands it.

Scalability plays a second role when thinking about really intelligent machines. "Singularity" evangelists (see the June 2008 issue of the IEEE Spectrum for an excellent pro-con discussion on this topic) base their positive expectations on Moore's Law and the daily experienced ever accelerating progress of technology. Ironically, also critics use the economy of scale as argument. Nothing in nature (energy, matter, etc.) can grow forever, and exponential growth of progress is – in their view – a subjective and potentially false perception. Even if our "truly intelligent" machines might not be the paradise host for our consciousness, once it moves out of its current medium, they might be attributed conscious attributes.

8 Social Implications

The relation between truly intelligent machines and mankind is subject of numerous science fiction stories, the murderous and tragic ones dominating. Isaac Asimov created a series of such stories, funny enough it is always man who fails and causes trouble, and never machines who are described honest and innocent. We should not underestimate the mental abilities of man to cope with this challenge when complex machines are all around. We can expect a pretty sober reaction by people that grow up with mass media and ubiquitous communication. These things will have their place in our world. More interesting is the question what place we will have in their world?

Generally we should distinguish between three different encounters that our society might have in future:

– Artifacts that are equal to animals
– Artifacts that are equal to humans (or at least appear like that)
– Artifacts that are superior to humans (post-singularity beings)

The first one can already be observed. Ranging from "tamagotchis" to Sony's Aibo robot-dog people bond to definitely inanimate tinboxes, more intelligent artificial animals will cause even more passionate behavior. Nothing wrong with that, if an artifact in deed has the intellect of a dog, why should it not receive the same respect as a real dog? Machines equal to us is more of a challenge. Phenomena like love, sympathy, racism or compassion might need a review if we live door-to-door with these creatures. And they might discuss the very same problems. The ultimate gain that we could get out of such a situation is to overcome all antique concepts of difference and realize that conscious beings are all equal. Anyway, we still did not encounter something entirely new. Throughout the history of our planet we met equal cultures and had to cope with it.

Meeting an outclassing entity, however, is something that no-body is prepared for. Antique mythology tells of gods which are equipped with extraordinary powers but are still helplessly extradited to very human emotions like love or anger. They can even be outsmarted sometimes, so they are by no means almighty. Modern religions paint a picture of a supreme creator, immune to our weaknesses, but captured in his own goodness and love for us. It is superior but not free. A mixture – an autonomous, superior being – threatens us. Would they exterminate us like we treat bugs and bacteria or would they explain us – finally – the true meaning of life? At least the question of social implications would not lay on our shoulders anymore, our superior creatures would ponder about that.

9   Conclusion and Outlook

Objects and devices get smarter every single day. It is a win-win situation for the users – who can get more services by machines – and manufacturers who always look for new products and markets. Their smartness today is shown in terms of function, usability, design, energy efficiency, sustainable product life cycle and the like. All of these are observable behavior, none of them can be directly assessed in terms of directly influences. With conscious machines on the other hand it would be possible to demand e.g. energy efficiency or security for children. Machine consciousness when developed following the template of human consciousness needs to show degrees of itself. It needs to be designed in the way it could potentially work in a real, living body in its mediator role between endogenous demands and the actual environment. These functions are elaborated by Freud and his successors who developed a functional model of the human mind. Another attribute of this model lies in the concept of the primary and secondary process and its implications for higher levels. Other humanities do not possess functional models or provide only behavioral models. However, psychoanalysis turned out to be applicable to surprisingly large extent into technical terms and in the future into systems.

The development of machine consciousness relys on many interdisciplinary findings presented above, whereby computer engineering and psychoanalysis will be the main contributors. Some of the requirements are already formulated in concepts, some still lack any idea for implementation. However, we tried to give the interested reader an impression of what has to be done to achieve machine consciousness in our view. We also have stressed the borders: that even very sophisticated solutions for human-like behaviour in terms of moving arms and feet, producing facial expressions, following speech dialogs, etc. do not contribute to making machines more conscious. The key for any stage of development of higher order mental content lies in subjective evaluation of lower level mental content. This is what children need to do from very soon after birth, and so conscious machines will have to. Some remakable capabilities resulting from this developmental process in children have been presented.

There is already a community, founded with the origination of the ENF – the first international engineering and neuro-psychoanalysis forum [40]. Many international researchers took notice of that event and even more than 100 came to attend. Things start to come together. We see this as radically new approach in AI. Many other approaches to reach human-like intelligence failed because of – in retrospect – clearly visible lacks in theory or methodology. This one is unique in terms of understanding and applying proven theories of human consciousness into machines. You are welcome to join us and enhance our efforts with your valuable ideas.

References

1. Baars, B.J.: Some essential differences between consciousness and attention, perception, and working memory. Consciousness and Cognition 6, 363–371 (1997)
2. Damasio, A.R.: The Feeling of What Happens: Body, Emotions and the Making of Consciousness. Econ Ullstein List Verlag, GmbH (1999)
3. Barnard, E., Palensky, B., Palensky, P.: Towards Learning 2.0. In: Proceedings of ICST IT-Revolutions 2008, Venice (2008)
4. Palensky, P., Lorenz, B., Clarici, A.: Cognitive and Affective Automation: Machines Using the Psychoanalytic Model of the Human Mind. In: Proceedings of First IEEE Engineering and Neuro-Psychoanalysis Forum, Vienna (2007)
5. Arlow, J.A., Brenner, C.: Psychoanalytic Concepts and the Structural Theory. International Universities Press, New York (1964)
6. Bateman, A., Holmes, J.: Introduction to Psychoanalysis – Contemporary Theory and Practice. Routledge, London (1995)
7. Chalmers, D.: The Puzzle of Conscious Experience. Scientific American, 62–68 (December 1995)
8. Chalmers, D.: Facing up to the problem of consciousness. Journal of Consciousness Studies 2(3), 200–219 (1995)
9. Dennett, D.: Who's on First? Heterophenomenology Explained. Journal of Consciousness Studies, Special Issue: Trusting the Subject (Part1) 10(9-10), 19–30 (2003)
10. Sandler, J., Holder, A., Dare, C., Dreher, A.U.: Freud's Models of the Mind. An Introduction. Karnac, London (1997)
11. Freud, S.: Formulations on the Two Principles of Mental Functioning. In: Strachey, J. (ed.&Trans.) The Standard Edition of the Complete Psychological Works of Sigmund Freud, vol. 12, pp. 218–226. Hogarth Press, London (1911)
12. Freud, S.: The Ego and the Id. Standard Edition XIX, 109–121 (1923)
13. Freud, S.: The Unconscious. Standard Edition XIV, 166–204 (1915)
14. Freud, S.: The Interpretation of Dreams. Standard Edition IV & V (1900)
15. Freud, S.: Three essays on the theory of sexuality. Standard Edition VII, 135–243 (1905)
16. Freud, S.: Beyond the pleasure principle. Standard Edition XVIII, 7–64 (1920)
17. Solms, M., Turnbull, O.: The Brain and the Inner World. Karnac, London (2002)
18. Solms, M.: Freud returns. Scientific American, 56–62 (May 2004)

19. Solms, M.: What is the "mind"? A neuro-psychoanalytical approach. In: Dietrich, D., Zucker, G., Bruckner, D., Fodor, G. (eds.) Simulating the mind, A technical, neuro-psychoanalytical approach, Springer, NewYork (2008)

20. Viola, P., Jones, M.: Rapid Object Detection using a Boosted Cascade of Simple Features. In: Conference on Computer Vision and Pattern Recognition (2001)

21. Mattern, F.: Ubiquitous Computing: Schlaue Alltagsgegenstände - Die Vision von der Informatisierung des Alltags. In: Bulletin EV/VSE, vol. 19, pp. 9–13 (2004)

22. Hainich, R.R.: The End of Hardware, A Novel approach to Augmented Reality. BookSurge Publishing (2006)

23. Lindwer, M., Marculescu, D., Basten, T., Zimmermann, R., Marculescu, R., Jung, S., Cantatore, E.: Ambient Intelligence Visions and Achievements; Linking abstract ideas to real-world concepts. In: Proceedings of the conference on Design, Automation and Test in Europe DATE 2003 (2003)

24. Endres, C., Butz, A., MacWilliams, A.: A Survey of Software Infrastructures and Frameworks for Ubiquitous Computing. Mobile Information Systems Journal 1(1) (2005)

25. Lagendijk, R.L.: The TU-Delft Research Program Ubiquitous Communications. In: Proceedings of the Twenty-first Symposium on Information Theory in the Benelux, pp. 33–44 (2000)

26. Roman, M., Hess, C.K., Cerqueira, R., Ranganathan, A., Campbell, R.H., Nahrstedt, K.: Gaia: A Middleware Infrastructure to Enable Active Spaces. IEEE Pervasive Computing, 74–83 (2002)

27. Mavrommati, I., Kameras, A.: The evolution of objects into Hyper-objects. Personal and Ubiquitous Computing 7(1), 176–181 (2003)

28. Gellersen, H.-W., Schmidt, A., Beigl, M.: Multi-Sensor Context-Awareness in Mobile Devices and Smart Artefacts. Mobile Networks and Applications 7, 341–351 (2002)

29. Lohse, M., Slusallek, P.: Middleware Support for Seamless Multimedia Home Entertainment for Mobile Users and Heterogeneous Environments, 217–222 (2003)

30. Want, R., Schilit, B., Adams, N., Gold, R., Petersen, K., Ellis, J., Goldberg, D., Weiser, M.: The PARCTAB ubiquitous computing experiment. In: Proceedings of the Fourth Workshop on Workstation Operating Systems (1995)

31. Bion, W.: A theory of thinking. International Journal of Psycho Analysis (40), 4 5 (1962)

32. Pfeifer, R., Scheier, C.: Understanding Intelligence. MIT Press, Cambridge (2001)

33. Brooks, R.: Intelligence without representation (47), 139–159 (1991)

34. Brooks, R.A.: I, Rodney Brooks, Am a Robot. IEEE Spectrum 06.08 (2008)

35. Alter, T.: Qualia. In: Nadel, L. (ed.) Encyclopedia of Cognitive Science, pp. 807–813. Macmillan Publishers Ltd., Basingstoke (2003)

36. Bringsjord, S.: The Zombie Attack on the Computational Conception of Mind. Philosophy and Phenomenological Research 59.1 (1997)

37. Holland, O (ed.): Machine Consciousness. Imprint Academic (2003)

38. Penrose, R.: The emperor's new mind. Oxford University Press, Oxford (1989)

39. Fonagy, P., Target, M., Gergely, G., Jurist, E.L.: Affect Regulation, Mentalization, and the Development of Self, 1st edn. Other Press (2000)

40. Dietrich, D., Fodor, G., Zucker, G., Bruckner, D. (eds.): Simulating the Mind. Springer, Heidelberg (2008)

41. von Förster, H.: Wissen und Gewissen: Versuch einer Brücke, 7th edn. (2006)

42. Yowell, Y.: Return of the zombie - Neuropsychoanalysis, consciousness, and the engineering of psychic functions. In: Dietrich, D., Zucker, G., Bruckner, D., Fodor, G. (eds.) Simulating the mind, A technical, neuro-psychoanalytical approach. Springer, NewYork (2008)

43. Beck, F., Eccles, J.C.: Quantum aspects of brain activity and the role of consciousness. In: Proceedings of the National Academy of Sciences of the United States of America, vol. 89, pp. 11357–11361 (1992)

44. Penrose, R., Hameroff, S.: Orchestrated objective reduction of quantum coherence in brain microtubules: The "orch OR" model for consciousness. Mathematics and Computers in Simulation 40, 453–480 (1996)

Data Mining on Distributed Medical Databases: Recent Trends and Future Directions

Yasemin Atilgan and Firat Dogan

Dogus University, Computer Engineering Department, Research Assistant
Acibadem, Istanbul, Turkey
{yatilgan,fdogan}@dogus.edu.tr

Abstract. As computerization in healthcare services increase, the amount of available digital data is growing at an unprecedented rate and as a result healthcare organizations are much more able to store data than to extract knowledge from it. Today the major challenge is to transform these data into useful information and knowledge. It is important for healthcare organizations to use stored data to improve quality while reducing cost. This paper first investigates the data mining applications on centralized medical databases, and how they are used for diagnostic and population health, then introduces distributed databases. The integration needs and issues of distributed medical databases are described. Finally the paper focuses on data mining studies on distributed medical databases.

Keywords: Data mining; Distributed Databases; Medical Databases.

1 Introduction

Today like all organizations healthcare organizations are also employing some sort of information systems. These systems provide technology to store data which take the form of numbers, text, chart or images. As a result large medical databases have grown in today's healthcare environment. These large medical databases store vast amount of data that is not used. It is an important issue in Information Age to extract knowledge from these data repositories, especially while healthcare organizations are facing a major challenge on improving service of quality delivered at affordable costs. The strategy to lower cost, raise quality and still be competitive in the information age is to build up strong healthcare information systems for knowledge management and decision support. Knowledge is the new capital of organizations. Today the economy is knowledge economy.

Using computer and information technologies in healthcare services helps to achieve efficiency, effectiveness in diagnostic decision making, cost economy, better risk management and strategic planning in a competitive healthcare environment [1].

Organizations are much more able to store data than to extract knowledge from it. Data modeling and analysis tools like data mining are widely used to generate knowledge rich environment. Using data mining techniques on medical data helps better decision making in diagnosis, better care on patients, finding way of preventing

M. Ulieru, P. Palensky, and R. Doursat (Eds.): IT Revolutions 2008, LNICST 11, pp. 216–224, 2009.

some diseases and early prevention in some medical areas, and all these provide cost reduction, better risk management and quality of services in healthcare services.

The improvements in information technologies have not only brought services to store data. Before the revolution in computer systems, most organizations had only a handful of computers, and for lack of a way to connect them, these operated independently from one another. The advances in microprocessors and networks changed that situation. The result of these technologies brought distributed systems in contrast to previous centralized systems [2]. Tanenbaum and Steen define distributed system as a collection of independent computers that appear to the users of the system as a single computer. Distributed systems made the connection of databases that are geographically or physically separate from each other possible. The developments in distributed technology also made information sharing possible. A shared information system can be considered as a series of computer systems - each is also a data repository - interconnected by some sort of communications network. Figure 1 visualizes a centralized database system and distributed database system.

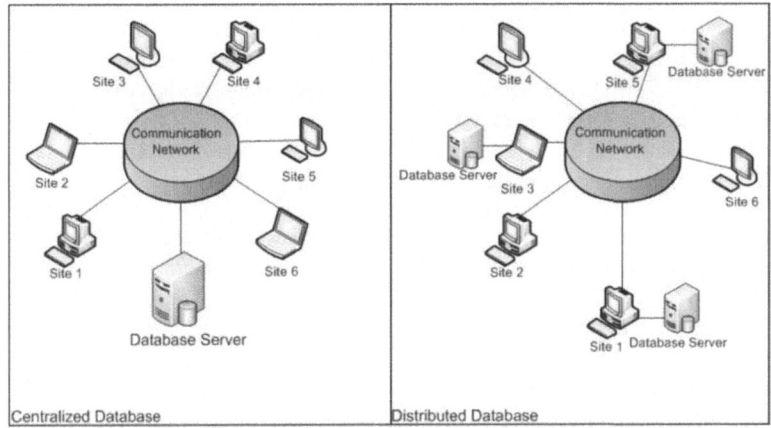

Fig. 1. Schematic representation of centralized and distributed databases

As a result of above mentioned developments in technology brings the concept of distributed data mining. The medical data is distributed because the data repositories are located in different hospital or different departments in one hospital and heterogeneous because patient's medical records can be stored in databases in different forms, inspecting results can be saved as imaging files, clinical cases or doctor's advices can be saved as text documents, and other video or images. Today the challenge is to extract knowledge from these distributed and heterogeneous data repositories. Using data mining techniques on data stored in centralized databases is common. It is possible to find applications of different data mining techniques on different kind of medical data, which is naturally stored in centralized databases or the data used has been centralized to accomplish the study. The application of data mining on distributed databases is still not a common practice.

The purpose of this paper is to review recent data mining trends on distributed databases, give an insight about the studies and applications, and finally explore the future directions in distributed medical databases.

The remainder of the paper is organized as follows. Section 2 provides a brief overview of data mining perspective on medical databases and data mining applications in this area; section 3 explores the researches and studies about data mining on distributed medical databases.

2   Data Mining on Medical Databases

Data mining is a field that uses database technology, statistics, pattern recognition, data visualization, machine learning and expert systems for knowledge discovery. A database is a collection of data that is organized so that its contents can easily be accessed, managed and updated [3]. In today's healthcare environment vast amount of clinical and personal data about patients are kept in medical databases which take the form of numbers, text, charts and images. Unfortunately, these data are rarely used to support clinical decision making. There is a wealth of information hidden in these data that is largely untapped [4]. The challenge is to transform these data into information and knowledge. Data mining tools make knowledge discovery possible from these large data repositories.

Frawley et al. define knowledge discovery as the nontrivial extraction of implicit, previously unknown, and potentially useful information from data:

"Given a set of facts (data) F, a language L, and some measure of certainty C, we define a pattern as a statement S in L that describes relationships among a subset FS of F with a certainty c, such that S is simpler (in some sense) than the enumeration of all facts in FS. A pattern that is interesting (according to a user-imposed interest measure) and certain enough (again according to the user's criteria) is called knowledge. The output of a program that monitors the set of facts in a database and produces patterns in this sense is discovered knowledge."

A lot of valuable knowledge hidden in the databases can be discovered using data mining approach, which is worth exploring. For example, to understand the relationships between characteristics of patient symptoms and the illness so that patients can utilize the results of this research to assist in guiding patients to connect their own symptoms to the type of illness accurately large data sets and classification methods in data mining technique can be used [6].   Applications of data mining techniques in health care systems vary from heart disease prediction [4] to early detection [7] or risk grouping [8] of prostate cancer, from predicting breast cancer survivability [9], to assessing healthcare resource utilization of lung cancer patients [10]. Kraft et al. studied for a process of knowledge generation for predicting the length of stay of spinal cord injury patients. They used nursing diagnosis to predict length of stay. As a result no one diagnostic cluster was found as the critical factor in the prediction model. They claim that the decision of data to be warehoused becomes very important for better application of data mining techniques and improve knowledge discovery. This also shows that data warehousing and the design issues in databases are also important, but this is a subject of another study. Chae et al.

presented analysis of healthcare quality indicators using data mining techniques. They identified the important factors influencing the inpatient mortality. They demonstrated how a data mining technique could be used in developing continuous quality improvement strategy. In literature it is possible to find other studies and applications implementing different data mining techniques using data from different areas in healthcare [13][14][15][16]. Some studies show that by accessing the right information, at right time at right place enables taking the right action on time or helps early prevention, while some studies emphasis the importance of quality and availability of data. Accessing the right knowledge provides diagnosing and treating patients, as well as to prevent and maintain some illnesses [17].

3 Distributed Medical Databases

3.1 Integration of Medical Databases

As healthcare institutions are hiring new information systems the data being digitalized is increasing exponentially. These information systems are classified according to place they are used and the type of data stored. Generally they are classified as Hospital Information Systems (HIS), Radiology Information Systems (RIS), Laboratory Information Systems (LIS), Picture Archiving and Communication System (PACS) and so on[18]. All these systems are centralized stand-alone systems, so they work independently from each other. As a result by the nature of the data type stored in these systems heterogeneity is common. The other issue in medical databases is, since the healthcare organizations are located in different places and have their own data repositories the medical data are stored in distributed databases by nature. So, distributed medical databases in this study refer to the medical databases that are distributed because of their nature. Designing a distributed database for a geographically dispersed organization is another issue.

Today, the healthcare systems around the world are moving towards the integration of these distributed data sources, to improve the efficiency in healthcare services through data sharing [19]. Building information systems without carefully planning and developing independent systems for new services, with medical data of a patient scattered in various stand-alone databases, causes cumulative inefficiencies in the information processing and sharing, thus leading to medical mishaps and potential risks in legal liabilities from patient care delivery [20]. As a result, integration of existing heterogeneous databases to provide effective and efficient knowledge discovery and better knowledge management in hospitals is one of the most popular issues in healthcare informatics.

Although database integration issues are not new to database researchers, hospitals have many unique characteristics which entail special integration design considerations for the following reasons [1]:

- Operational efficiency
- Cost economy
- Effective diagnostic decision making
- Emergency care services
- Legal requirements

- Strategic planning and risk management
- Education and research

They proposed a virtual database integration approach for the database integration needed in a hospital to solve data redundancy, data inconsistency, data model incompatibility, time delay and inconvenience problems and satisfy the information handling needs from various hospital services and functions and also retain the autonomy of departmental system development and provide central planning and control mechanism for system expansion. Integration is also important for multi-site healthcare organizations. These organizations implement heterogeneous information management systems interacting with distributed databases [21]. The study explores the 'middleware services' as a solution to guarantee data exchange across different types of applications and database management systems, and reduce the costs of systems development and modification. Their study concluded that their design architecture is a highly effective tool for developing reusable object/components that can substantially reduce the time and effort required for system development and maintenance. In [22] the integration problem of heterogeneous data in medicine is studied. They present a concept and solution to support intra and also inter-institutional integration of systems that are being used in healthcare institutions. Their concept was based on requirements of researchers and clinicians to integrate data from various existing component systems. They represented a "light-weight" approach to interconnect existent data sources, because of legal and regulatory restrictions. Another study about integration is given in [23]. They developed various methods and tools for database integration, because they mention that different biological and clinical research projects are based on collaborative efforts among international organizations. The method they present detects the inconsistencies automatically and stores the corresponding transformations in a formal structure to support knowledge discovery professionals. In literature it is possible to find different kinds of integration solutions like a framework used to integrate molecular biology database using context graph [24], an architecture based on multi-agent system and grid technology [25] and other studies for the integration of genomic data [26][27]. Integration of medical databases is an important issue and necessary for knowledge discovery in databases. Even it is not discussed in this study, also legal and privacy issues are important and should be considered during integration process.

3.2 Data Mining Applications on Distributed Medical Databases

Since modern medicine generates almost daily huge amounts of heterogeneous data and the stored medical data may contain images, signals, clinical information, cholesterol levels, etc., as well as the physician's interpretation [28], researchers present architecture for cooperative work in heterogeneous medical information, such as Hospital Information System(HIS) and Laboratory Information System (LIS) [29].

A methodology and operational framework for applying data mining techniques from distributed and heterogeneous clinical data sources is presented by [30]. In their study they state that epidemiological studies are important for health prevention and health prevention is highly dependent on information transfer. To solve the problem of mining heterogeneous and distributed data sources they indicate that a multi-phase data integration procedure should be followed. After the integration methodology, they propose their study on the specifics of knowledge discovery processes. They particularly study on discovery of interesting associations between the recorded patients' clinical data items. Association analysis is the discovery of association rules showing attribute-value conditions that occur frequently together in a given set of data [31]. They make the assumption that these associations may be linked with indicative epidemiological and health-indicators. Their results show the effectiveness of respective distributed data mining operations. As a result they discover and form interesting associations according to specific query posted by the user via the patient clinical data directory service. Another study presents an intelligent agent based framework for knowledge discovery comprising multiple heterogeneous healthcare data resources [32]. They argue that with the existence of multiple heterogeneous data repositories in a healthcare enterprise a distributed data community, such that any data mining effort draws upon the 'holistic' data available within the entire healthcare enterprise should be established. The proposed multi Agent-Based Data Mining Info-Structure (ADMI), uses the advantage of a multi-agent architecture which features the amalgamation of various types of intelligent agents, each responsible for an independent task. It is responsible for the generation of data-mediated diagnostic-support and strategic services. They designed the Interface Agent (IA) to collect user-specification for a data mining service via a web-based interface, Data Collection Agent (DCA) to facilitate the on-demand retrieval of relevant data from the multiple healthcare data repositories, Data Mining Agent (DMA) for coordinating the entire data mining activities, and Services Generation Agent (SGA) to process the data mining results produced by DMA to generate decision-support or strategic services as per the user's request. They demonstrate that autonomous, reactive and proactive intelligent agents provide an opportunity to generate end-user oriented, packaged, value-added decision-support/strategic planning services for healthcare professionals and managers. Their work was leverage and a prototype version of their framework is under development.

As the volume of data stored increase, new challenges for their effective understanding raise. Since the processes and activities that are computationally intensive, collaborative and distributed in nature are involved in knowledge discovery in large data repositories, high level frame works like Knowledge Grid are developed [33]. Grid computing aims to aggregate distributed computing resources, hide their specifications and present a homogenous interface to end users for high performance or high throughput computation [34]. The grid can play a significant role in providing an effective computational infrastructure support for data mining from very large datasets maintained over geographically distributed sites by using computational power of distributed systems [35].

Grid technology can be used in hospitals and medical related works to share and integrate heterogeneous medical sources [36]. The study also introduces the related works on medical grid applications. The applications introduced are mostly related with databases storing medical images like CT, MRI and mammograms. Since the digital medical images represent a tremendous amount of data, a hospital is capable of producing several Terabytes of medical image data each year. Because of this reason using grid technologies for medical image databases is more common.

The DataMiningGrid system is a system that has been designed and developed in order to meet the requirements of modern and distributed data mining scenarios. The system was recently built on top of existing Globus technology inter alia to address the requirements of a community of medical users and enable them to perform on-the-fly analysis of geographically distributed medical databases. DataMiningGrid system provides tools and services facilitating the grid-enabling of various applications including data mining and statistical applications without major intervention on the application side [37]. The details of DataMiningGrid system architecture is introduced with details in [38]. The study also presents the grid-enabling data mining applications and the requirements. It is claimed that there is still a long way to go before grids can be widely used in the medical domain, however prototypes and experiments are developed on medical applications [39].

4 Conclusion

The studies show that there is variety of data mining applications on medical database. These studies have been conducted on different subjects in healthcare like cancer prediction, length of hospital stay optimization and so on. In literature different data mining techniques applied to centralized information repositories, and some studies make the comparison between these techniques. We can say that there is a lack of data mining applications on distributed medical databases. The recent trends on data mining on distributed medical databases are mostly in design base, or applied in sample databases. The future directions on database and data mining applications is designing more scalable and internet based data mining architectures so that the data will be available anytime by everyone who has access to Internet.

References

1. Liu Sheng, O.R., Garcia, H.-M.C.: Information Management in Hospitals: An Integrating Approach. In: 9th IEEE International Phoenix Conference on Computers and Communications, pp. 296–303. IEEE press, Scottsdale (1990)
2. Tanenbaum, A.S., Steen, M.: Distributed Systems: Principles and Paradigms, pp. 2–3. Prentice Hall, New Jersey (2002)
3. Obenshain, M.K.: Application of Data Mining Techniques to Healthcare Data. Infection Control and Hospital Epidemiology, 690–695 (2004)
4. Palaniappan, S., Awang, R.: Intelligent Heart Disease Prediction System Using Data Mining Techniques. In: IEEE/ACS International Conference on Computer Systems and Applications, pp. 108–115. IEEE press, Doha (2008)
5. Frawley, W.J., Piatetsky-Shapiro, G., Matheus, C.J.: Knowledge Discovery in Databases: An Ovrview. AI Magazine, 57–70 (1992)
6. Chang, C.L.: A Study of Applying Data Mining to Early Intervention for Developmentally-delayed Children. Expert Systems with Applications, 407–412 (2007)
7. Zhang, Z., Zhang, H.: Development of a Neural Network Derived Index for Early Detection of Prostate Cancer. IEEE International Joint Conference on Neural Networks, 3636–3641 (1999)
8. Churilv, L., Bagirov, A.M., Schwartz, D., Smith, K., Dally, M.: Improving Risk Grouping Rules for Prostate Cancer Patients with Optimization. In: IEEE Proceedings of the International Conference on System Sciences, Hawai (2004)

9. Delen, D., Walker, G., Kadam, A.: Predicting Breast Cancer Survivability: a comparison of three data mining methods. Artifical Intelligence in Medicine, 113–127 (2005)

10. Phillips-Wren, G., Sharkey, P., Dy, S.M.: Mining Lung Cancer Data to Assess Healthcare Resource Utilization. Expert Systems with Applications, 1611–1619 (2008)

11. Kraft, M.R., Desouza, K.C., Anndrowich, I.: Data Mining in Healthcare Information Systems: Case Study of a Veterans' Administration Spinal Cord Injury Population. In: IEEE Proceedings of the International Conference on System Sciences, Hawaii (2002)

12. Chae, Y.M., Kim, H.S., Tark, K.C., Park, H.J., Ho, S.H.: Analysis of Healthcare Quality Indicator Using data Mining and Dceision Support System. Expert Systems with Applications, 167–172 (2003)

13. Lee, S., Abbott, P.A.: Bayesian Networks for Knowledge Discovery in Large Datasets: Basics for Nurse Researchers. J. Biomedical Informatics, 389–399 (2003)

14. Lin, F., Chou, S., Pan, S., Chen, Y.: Mining Time Dependency Patterns in Clinical Pathways. In: IEEE Proceedings of the International Conference on System Sciences, Hawaii (2000)

15. Wilson, A.M., Thabane, L., Holbrook, A.: Application of Data Mining Techniques in Pharmacovigilance. British Journal of Clinical Pharmacology, 127–134 (2003)

16. Silva, A., Cortez, P., Santos, M.F., Gomes, L., Neves, J.: Mortality Assessment in Intensive Care Units via Adverse events Using Artificial Neural Networks. Artificial Intelligence in Medicine, 223–234 (2006)

17. Goodwin, L., VanDyne, M., Lin, S., Talbert, S.: Data Mining Issues and Opportunities for Building Nursing Knowledge. J. Biomedical Informatics, 379–388 (2003)

18. Ahn, C., Nah, Y., Park, S., Kim, J.: An integrated medical information system using XML. In: Kim, W., Ling, T.-W., Lee, Y.-J., Park, S.-S. (eds.) Human Society Internet 2001, vol. 2105, pp. 307–322. Springer, Heidelberg (2001)

19. Au, R., Croll, P.: Consumer-Centric and Privacy-preserving Identity Management for Distributed e-Health Systems. In: IEEE Proceedings of the International Conference on System Sciences, pp. 1–10 (2008)

20. Troyer, G.T., Salman, S.L.: Handbook of Health Care Risk Management. Aspen Systems Corporation, Maryland (1986)

21. Chu, S., Cesnik, B.: A three-tire Clinical Information Systems Design Model. International J. Medical Informatics, 91–107 (2000)

22. Wurst, S.H.R., Lamla, G., Schlundt, J., Karlsen, R., Kuhn, K.A.: A Service-oriented Architectural Framework for the Integration of Information Systems in Clinical Research. In: IEEE Proceedings of the International Symposium on Computer-Based Medical Systems, pp. 16–163 (2008)

23. Anguita, A., Perez-Ray, D., Crespo, J., Mojo, V.: Automatic Generation of Integration and Preprocessing Ontologies for Biomedical Sources in a Distributed Scenario. In: IEEE Proceedings of the International Symposium on Computer-Based Medical Systems, pp. 336–341 (2008)

24. Khan, N., Rahman, S., Stockman, A.G.: A Framework for Molecular Biology Database Integration Using Context Graph. In: IEEE Proceedings of the International Symposium on Computer-Based Medical Systems, pp. 21–26 (2004)

25. Di Lecce, V., Amato, A., Calabrese, M.: Data Integration in Distributed Medical Information Systems. In: Canadian Conference on Electrical and Computer Engineering, pp. 1497–1502 (2008)

26. Gros, P.E., Herisson, J., Ferey, N., Gherbi, R.: Combining Applications and Databases Integration Approaches in a Common Distributed Genomic Platform. In: IEEE Proceedings of the International Conference on Advanced Information Networking and Applications, pp. 433–438 (2005)

27. Douthart, R.J., Pelkey, J.E., Thomas, G.S.: Database Integration and Visualization of Maps of the Human Genome Using the GnomeView Interface. In: IEEE Proceedings of the International Conference on System Sciences, pp. 49–57 (1994)
28. Cios, K.J.: Medical Data Mining and Knowledge Discovery. Studies in Fuzziness and Soft Computing. Physica - Verlag (2001)
29. Li, K., Yao, D.: Cooperative Work in Heterogeneous Medical Information Systems. In: IEEE Proceedings of the International Conference on Communications, Circuits and Systems (2006)
30. Potamias, G.A., Moustakis, V.S.: Knowledge Discovery from Distributed Clinical Data Sources: The Era for Internet-based Epidemiology. In: IEEE Proceedings of EMBS International Conference, pp. 3638–3641 (2001)
31. Han, J., Kamber, M.: Data Mining Concepts and Techniques. Morgan Kaufmann, US (2001)
32. Zaidi, S.Z.H., Abidi, S.S.R., Manickam, S.: Distributed Data Mining from Heterogeneous Healthcare Data Repositories: Towards an Intelligent Agent-Based Framework. In: IEEE Proceedings of Symposium on Computer-Based Medical Systems (2002)
33. Congiusta, A., Talia, D., Trunfilo, P.: Distributed Data Mining Sevices Leveraging WSRF. Future Generation Computer Systems, 34–41 (2007)
34. Luo, P., Lü, K., Shi, Z., He, Q.: Distributed Data Mining in Grid Computing Environments. Future Genertion Computer Systems, 84–91 (2007)
35. Luo, J., Wangc, M., Hud, J., Shia, Z.: Distributed Data Mining on Agent Grid: Issues, platform and development kit. Future Generation Computer Systems 3, 61–68 (2007)
36. Zheng, R., Jin, H., Zhang, Q., Liu, Y., Chu, P.: Heterogeneous Medical Data Share and Integration on Grid. In: IEEE Proceedings of the International Conference on BioMedical Engineering and Informatics, pp. 905–909 (2008)
37. Jarm, T., Kramar, P., Županič, A. (eds.) : Medicon 2007. IFMBE Proceedings 16, 166–169 (2007)
38. Stankovski, V., Swain, M., Kravtsov, V., Niessesn, T., Wegener, D., Kindermann, J., Dubitzky, W.: Grid-enabled Data Mining Applications with DataMiningGrid: An architectural perspective. Future Generation Computer Systems, 1–21 (2007)
39. Montagnat, J., Breton, V., Magnin, I.E.: Using grid Technologies to Face Medical Image Analysis Challenges. In: IEEE/ACM 3rd International Symposium on Cluster Computing and the Grid, pp. 1–5 (2003)

Economic Activity and Climate Change in a Structural Framework: A First Approach

Panayotis Michaelides and Kostas Theologou

National Technical University of Athens
School of Applied Mathematics & Physical Sciences
Department of Humanities, Social Sciences and Law
9 Heroon Polytechneiou, Zografou Campus 15780, Athens, Greece
{pmichael,cstheol}@central.ntua.gr

Abstract. The considerable increases in greenhouse gases emissions and the subsequent climate changes are directly associated with the current rates of economic growth. As a result, the increasing environmental problems should result in an effort to accurately measure and control climate changes. In the present paper, we propose a theoretical inter-industry model for analyzing how economic activity, by industrial sector, affects the climate. We believe that the proposed methodology could be utilized for the feedback of the policy formulation procedure and could provide a vehicle for expanding conventional climate change analysis in economics.

Keywords: socioeconomic sphere, environment, climate change, economic growth, input-output.

1 Introduction

Nowadays, the globe is facing numerous challenges: on the one hand there are the problems related to poverty and socioeconomic instability, whereas, on other hand, there are the problems related to the environment and the changing climate. The environmental crisis has resulted in phenomena such as high temperatures, melting of glaciers, floods, rise in sea levels, droughts, water contamination, deforestation, loss of biodiversity, and so on. These problems are, in turn, associated with the unsustainable growth and increasing consumption patterns. In this context, the relation between the socioeconomic sphere and the environment has attracted attention in the literature and sustainable development has become a 'hot' topic in the research agenda of many countries.

For instance, as it was argued by Ayres and Simonis [1] and Bringezu [2] the interaction of the socioeconomic sphere with the environment is closely linked to the amount of material, emissions and energy expanded. As a result, the increasing environmental problems have resulted in an effort to accurately measure and control the climate changes in most developed and developing countries.

Meanwhile, climate change experts are alarmed by the considerable increase in greenhouse gases emissions that are associated with the current (economic) growth rates, and they tend to recommend strong mitigation action. In this framework, the question of climate change is a very important issue and many countries are required

M. Ulieru, P. Palensky, and R. Doursat (Eds.): IT Revolutions 2008, LNICST 11, pp. 225–231, 2009.
© ICST Institute for Computer Sciences, Social-Informatics and Telecommunications Engineering 2009

to implement new directives, which set long-term quality objectives introduced by the Kyoto Protocol. Of course, several industrial countries worry that economic growth might be harmed if they engage in mitigation.

Put concretely, the main problem that policymakers are, nowadays, facing is to analyze the implications for climate change of several policy trajectories with a view to finding a set of actions that minimize the impact of economic growth on climate change. In the present paper, we are taking one step towards achieving this goal by focusing on a theoretical model for analyzing how economic activity by industrial sector affects the climate.

The paper is organized as follows: section 2 analyzes the relation between energy use, climate change and economics by considering previous studies; section 3 sets out the proposed theoretical model; finally, section 4 concludes.

2 Energy Use, Climate Change and Economic Analysis

Human civilization was built on three pillars: (a) the ability of man to grow enough food to adequately support the population that is engaged in tasks other than growing food; (b) the ability to live in large communities (groups) that are able to support major societal and administrational institutions such as the parliamentary system, the judiciary system and the educational system; (c) the ability to 'exploit' labour in the broader sense and exchange its products. In other words, these three pillars express Society, Politics and Economics.

The march towards a fossil-fuel-dependent economy started as early as in the England of Edward I, for he detested the smell of coal that he banned (in 1306) the burning of coal. In spite of the King's prohibition, eventually the English became the first Europeans to burn coal on a large scale [3]. However, by 1700 a thousand tones per day were being burned in the city of London. An 'energy crisis' soon emerged. England's mines had been dug so deep that they were filled with water. A way of pumping it out on the surface had to be found. The man who discovered how this could be effectuated was Thomas Newcomen (1664-1729); his device burned coal to produce steam, which was then condensed to create a vacuum that moved a piston which pumped the water. The first Newcomen engine was installed in a Staffordshire coal mine in 1712. Fifty years later, hundreds of them were at work in mines across the nation, and England's coal production had grown to 6 million tones per year [4].

The ingenious Scottish James Watt (1736-1819) improved on Newcomen's design and in 1784 William Murdoch (1754-1839) produced the first mobile steam engine. From that time on, the coming 19th century was to be the century of coal. No other fuel source was used for industrial purposes, transportation, heating or cooking. In 1882 Thomas Edison (1847-1931) inaugurated the world's first electric light power station in Manhattan.

In the aftermath of Word War I, when people stoked the coal and wood fuelled stoves, great amounts of CO_2 were released, which are still warming Earth today. Most of the damage was done after 1950 when people kept driving and moving around in their vehicles and kept powering their manual work saving household appliances from inefficient coal-burning energy stations [5]. Most to blame is the baby-boomer generation born in the years following World War II. Half of the energy amounts generated since the Industrial Revolution have been consumed in the last twenty years.

Nowadays, the study of greenhouse gases, global warming and climate change has become a 'hot' topic in the literature. The greenhouse gases trap heat near the surface of the planet. As they increase, this 'trapped' heat leads to global warming. The warming destabilizes the planet's climate and potentially leads to climate change. Climate is regarded as the sum of all weathers over a certain period of time, for a region or for the planet as a whole.

Economy fosters climate change in all three stages of its activity, i.e. production, in distribution and consumption. Production needs energy. Supplying energy is a very profitable activity. In the developed countries, energy use grows at the rate of approx. 2% per annum. With such low rates of growth the only way for one sector to expand is to take from another sector's share.

Distribution also creates pollution, approximately about a third of global carbon dioxide (CO_2) emissions. Consumption is the positive feedback of production; the demand for more products fosters not only technological innovations in business and economic activities but also burdens the planet with acceleration in climate changes; humans are already consuming more of the planet's resources than is sustainable.

Based on the information on the average surface temperature of Earth from AD 1000 to 2100 we notice that prior to 1900 the average temperature was significantly lower and equal to 13.7^0C ([6] - [9]). Also, the nine out of ten warmest years ever recorded have occurred after 1990.

Since the great explosion of energy consuming machines (i.e. during and after the Industrial Revolution) the fossil-fuelled economy has been nourished principally by coal (19th c), oil (20th c) and gas (21st c) ([10]-[11]). The natural gas resources seem to be sufficient to cover the needs of our economy until 2050. The latter, methane, is the fossil fuel with the least carbon content. So, from a climate change perspective there is a vast difference between burning gas or coal. The energy that is liberated when we burn these fuels comes from carbon and hydrogen. In simple words, because carbon causes climate change, the more carbon-rich a fuel is, the greater its environmental load and the greater climate changes it creates.

The literature on the modeling of climate change in economic analysis is seriously limited. Despite the fact that the theoretical and empirical literature on economic growth addresses many issues that are relevant to the climate change problem it rarely discusses climate change explicitly.

More precisely, most climate change models focus on estimating the cost of mitigation. A few papers use growth models (e.g. [12]) and analyze the consequences of mitigation for growth, but they do not examine how the structure of the underlying growth model constrains the results. In addition, most climate change models analyze the problem in a cost-efficiency framework, and hence do not explicitly model damages (e.g. [13]–[14]). A few climate change models analyze the problem in a cost-benefit framework and explicitly model damages (e.g. [15]), but they usually rely on simplified growth models that are insufficient to address the structural complexity of the relationship between impacts of economic growth and climate change that would be of interest to policymakers.

Undoubtedly, the *Input-Output (IO)* framework is a very powerful tool for environmental analysis ([16]–[17]). Also, *Life Cycle Analysis* (LCA) has been widely adopted during the last decade [18] and there have been some attempts to make use of input-output analysis in studying *Product Life Cycles* (PLC) [19]. Meanwhile, many economists have cast the problem of environmental pollution and climate change in terms of the *Cost-Benefit Analysis* (CBA) framework ([20]–[21]) or even in *neoclassical optimal control* models (e.g. [12]) which are usually based on a Ramsey-type model of economic growth. Finally, the study of climate change and environmental impacts associated with household consumption has also been an area of considerable attention for the last years. A few examples include works looking at the broader range of impacts associated with *Life Style Models* (LSM) in general ([22]–[23]).

3 The Theoretical Model

As we have seen, several studies use the input-output framework for estimating the total emission coefficient per unit of production in each sector of economic activity and for the final demand categories (e.g. [24]-[27]). In this paper, the input-output approach is used in order to calculate *climate change multipliers* for the various industries and expressions of climate characteristics.

The proposed approach recognizes the fact that environmental pollution and the subsequent climate change proceeds at varying rates in different sectors, a fact which is very important for comparison among different countries since the intensive competition in the world market is shaped at the industry level [28].

The proposed approach will be able to identify the (total) climate change in the environment that will be generated in case of a change in an industry's final demand. More precisely, the proposed model answers the following question:

A change in an industry's final demand, what change (as expressed in measurable climate characteristics) will generate in the climate?

The methodology for constructing embodied *climate change indicators* could be built as follows [27]:

Let Z be the matrix of intermediate deliveries, X the vector of gross outputs, and Y vector of the final demands. The input coefficients matrix is obtained as follows:

$$A = Z x^{-1} \tag{1}$$

where $x = diagX$ denotes the diagonal matrix whose elements consist of vector X.

The balance equations can be written as:

$$X = AX + Y. \tag{2}$$

Solving the balance equation for X, we obtain:

$$X = (I - A)^{-1} Y. \tag{3}$$

where $(I - A)^{-1}$ denotes the Leontief inverse.

For each industry i, we then define the direct intensity coefficients (a_{ik}) of climate change as quantities expressing climate change (e.g. cm, kgr, etc) (E_{ik}) per gross output ($\mathbf{Xi}$), where k denotes the type of climate characteristic (e.g. rain, snow, ice, wind, etc):

$$a_{ik} = E_{ik} / Xi \quad (i = 1,2,\ldots,n) \ . \tag{4}$$

The next step is to calculate the total induced emission, including the direct effects, as follows:

$$e_{ik}' = a_{ik}'(I - A)^{-1} \ . \tag{5}$$

where an accent denotes transposition.

An critical issue that remains to be addressed is the (technical) construction of the relevant $\mathbf{E_{ik}}$ matrix. It could be based on NAMEA (National Accounting Matrix Including Environmental Accounts) tables which consist of a conventional national accounting matrix extended to include environmental accounts in physical units. Apparently, this highly technical issue, which is obviously extraneous to the scope of our paper, provides an excellent example for future research.

4 Conclusions

Nowadays, the relation between the socioeconomic sphere and the environment has attracted increasing attention given that the world economy is facing numerous challenges related to the environment and the changing climate caused by the unsustainable growth and increasing consumption patterns. As a result, the increasing environmental problems should result in an effort to accurately measure and control the climate changes since the considerable increase in greenhouse gases emissions and the subsequent climate changes are directly associated with the current (economic) growth rates. In this context, the main problem that policymakers are, nowadays, facing is to analyze the implications for climate change of several policy trajectories with a view to minimizing the impact of economic growth on climate change. In the present paper, we propose a theoretical model for analyzing how economic activity by industrial sector affects the climate. We believe that the proposed methodology can be utilized for the feedback of the policy formulation procedure and that with its generality, conformity with theory and simplicity of structure will provide a vehicle for expanding and improving conventional climate change analysis in economics. Some technical issues which remain to be addressed are fine examples for future research in the field.

References

1. Ayres, R.U., Simonis, U.E. (eds.): Industrial Metabolism: Restructuring for Sustainable Development. United Nations University Press, Tokyo (1994)
2. Bringezu, S.: Towards sustainable resource management in the European Union. Wuppertal Papers No. 121, Wuppertal Institute for Climate, Environment and Energy (2002)
3. Freese, B.: Coal: A Human History. The Perseus Books Group, Cambridge (2002)

4. Pacey, A.: Technology in World Civilization; a thousand-year history. Blackwell, Oxford (1990)
5. Flannery, T.: We Are the Weather Makers. In: The Story of Global Warming. The Text Publishing Co., Melbourne (2006)
6. Mann, M.E., Bradley, R.S., Hughes, M.K.: Global-scale temperature patterns and climate forcing over the past six centuries. Nature 392, 779–787 (1998)
7. Mann, M.E., Cane, M.A., Zebiak, S.E., Clement, A.: Volcanic and Solar Forcing of the Tropical Pacific over the Past 1000 Years. Journal of Climate 18, 447–456 (2005)
8. Jones, P.D., Mann, M.E.: Climate over Past Millennia. J. Reviews of Geophysics (RG 2002) 42 (2004) doi: 10.1029/2003RG000143
9. McIntyre, S., McKitrick, R.: The IPCC, the "Hockey Stick" Curve, and the Illusion of Experience: Re-evaluation of Data Raises Significant Questions. In: Washington Roundtable on Science & Public Policy, November 18, pp. 3–16. The George C. Marshall Institute, Washington (2003)
10. Yeomans, M.: Oil: Anatomy of an Industry. New Press, New York (2004)
11. Yergin, D.: The prize: the Epic Quest for Oil, Money, and Power. Simon & Schuster, New York (1991)
12. Nordhaus, W.: Rolling the DICE: An Optimal Transition Path for Controlling Greenhouse Gases. Cowles Foundation Discussion Paper No 1019. Yale University, New Haven (1992)
13. McKibbin, W.J., Wilcoxen, P.J.: The theoretical and empirical structure of the GCubed model. Economic Modelling 16, 123–148 (1999)
14. Paltsev, S., Reilly, J.M., Jacoby, H.D., Eckaus, R.S., McFarland, J., Sarofim, M., Asadoorian, M., Babiker, M.: The MIT Emissions Prediction and Policy Analysis (EPPA) Model: Version 4. MIT Global Change Joint Program Report 125. MIT Press, Cambridge (2005)
15. Manne, A., Richels, R.: Buying greenhouse insurance: the economic costs of carbon dioxide emission limits. MIT Press, Cambridge (1992)
16. Leontief, W.: Environmental Repercussions and the Economic Structure: An Input-Output Approach. Review of Economics and Statistics 52(3), 262–271 (1970)
17. Miller, R.E., Blair, P.D.: Input-Output Analysis. Prentice-Hall, Inc., Englewood Cliffs (1985)
18. Bras, B.: Incorporating Environmental Issues in Product Design and Realization. UNEP Industry and Environment 20(1-2), 5–13 (1997)
19. Borland, N., Kaufmann, H.P., Wallace, D.: Integrating Environmental Impact Assessment into Product Design: A Collaborative Modeling Approach. In: Paper of the 1998 ASME Design Technical Conference (1998)
20. Peck, S., Teisberg, T.: CETA: a model for carbon trajectory assessment. The Energy Journal 13(1), 1–21 (1992)
21. Cline, W.: The Economics of Global Warming. Institute of International Economics, Washington (1992)
22. Tukker, A., Jansen, B.: Enviromental Impacts of Products – A Detailed Review of Studies. Journal of Industrial Ecology 10, 159–182 (2006)
23. Gutowski, T., Taplett, A., Allen, A., Banzaert, A., Cirinciore, R., Cleaver, C., Figueredo, S., Fredholm, S., Gallant, B., Jones, A., Krones, J., Kudrowitz, B., Lin, C., Morales, A., Quinn, D., Roberts, M., Scaringe, R., Studley, T., Sukkasi, S., Tomczak, M., Vechakul, J., Wolf, M.: Environmental Life Style Analysis (ELSA). In: IEEE International Symposium on Electronics and the Environment, San Francisco, USA, May 19-20 (2008)

24. Wier, M.: Sources of Changes in Emissions from Energy. Economic Systems Research 10, 99–112 (1998)
25. Ostblom, G.: The environmental outcome of emissions-intensive economic growth: a critical look at official growth projections of Sweden up to year 2000. Economic System Research 10, 19–29 (1998)
26. Wilting, H.C., Biesiot, W., Moll, H.C.: Analysing Potentials for Reducing the Energy Requirements of Households in the Netherlands. Economic Systems Research 11, 233–243 (1999)
27. Braibant, M.: Environmental Accounts in France: NAMEA pilot study for France. In: Fourteenth International Conference on Input-Output Techniques (2002)
28. European Commission: EU Sectoral Competitiveness Indicators. Office for the Official Publications of the European Communities, Luxembourg (2005)

Towards Ontology as Knowledge Representation for Intellectual Capital Measurement

B. Zadjabbari, P. Wongthongtham, and T.S. Dillon

Digital Ecosystems and Business Intelligence Institute,
Curtin University, WA, Australia
behrang.zadjabba@postgrad.curtin.edu.au,
P.Wongthongtham@cbs.curtin.edu.au,
t.dillon@curtin.edu.au

Abstract. For many years, physical asset indicators were the main evidence of an organization's successful performance. However, the situation has changed after information technology revolution in the knowledge-based economy. Since 1980's business performance has not been limited only to physical assets instead intellectual capital are increasingly playing a major role in business performance. In this paper, we utilize ontology as a tool for knowledge representation in the domain of intellectual capital measurement. The ontology classifies ways of intangible capital measurement.

Keywords: Intellectual capital measurement, market capital, social capital, human capital, ontology knowledge representation.

1 Introduction

One of the main responsibilities of managers in an organization is to make the organization more productive. For many years, physical asset indicators were the main evidence of an organization's successful performance. However, the situation has changed after information technology revolution in the knowledge-based economy. Since 1980's business performance has not been limited only to physical assets instead the different types of intellectual capital are going to play a main role in business performance. Human capital was one of the first intellectual capital that business owners understood its importance to achieve business goals. In 90's decade, business owners were focused more on customer value and due to the fast growth of communication tools, communication within organization between employees and between customers, knowledge sharing between market components is now one of the most important assets in knowledge-based economy. In knowledge based economy, knowledge is core competency and key competitive advantage for business. This knowledge comes from internal resource data or external resource data. Also knowledge validity and trust between agents of the business like customers-to- customers, employees-to-employees, employers-to-employees, employers-to- customers are the main parts of the modern

M. Ulieru, P. Palensky, and R. Doursat (Eds.): IT Revolutions 2008, LNICST 11, pp. 232–239, 2009.

business environment. Sustainable business performance requires an effective system including all data resources and communication between these resources.

Feedback and business performance measurement is necessary to evaluate business and improve performance. Managers need to have a good view of their organisation for their short term planning and / or strategic planning in which the business performance methods can provide them. After the 1980s different measurement methods have been presented which focus on intellectual capital such as Balanced Scorecard method (BSC) [1], Skandia navigator model [1], Investor assigned market value [1], etc. All of these methods have mentioned different kinds of intangible assets and tried to measure them in different ways such as proxy measures, checklists, surveys, etc. Additionally, most of these methods focused on internal-based data and static data however the environment is very dynamic thus dynamic based measurement tools are needed.

In this paper, we propose an ontology as knowledge representation methodology for intellectual capital measurement. We utilize an ontology as a tool for knowledge representation in domain of intellectual capital measurement. It contributes to the following knowledge, experience and competence management tasks:

- Offer a sensible measurement tool to assess intellectual capital;
- A balance between different capital can be made within business activities to create business sustainability in future;
- Have a consensus view of intellectual capital measurement.

The ontology classifies ways of intangible capital measurement. We identify three components that constitute the concept of intellectual capital: social capital, market capital, and human capital. In the next sections, we discuss in detail about social capital, market capital, and human capital measurement. We then present the measurement in an ontology as knowledge representation of intellectual capital measurement. We conclude the work in the last section.

2 Social Capital Measurement

Social capital now plays an important role in economic development. Social capital is one of the key factors in an organization's success. It is a challenge to find suitable tools to measure the level of social capital. Fukuyama describes social capital as an ability of people to work together for common purposes in groups or organizations [2,3]. Putnam indicates that social capital is the norms and networks of civil society that enable groups of individuals to co-operate for mutual benefit (and perhaps for broader social benefit) and may allow social institutions to perform more productively [4]. Putnam also points out that social capital is embodied in forms such as civic and religious groups, bonds of family, informal community networks, kinship and friendship, and norms of reciprocity, volunteerism, altruism and trust [5]. From Deardorff's Glossary of International Economics, social capital is the network of relationships among persons, firms, and institutions in a society, together with the associated norms of behavior, trust, cooperation, etc., that enables a society to function effectively [6].

As it comes to the definition of social capital, the social capital is related to people's willingness to make connections and density of the information that is transmitted in those connections. Also transmitted information has different influence and it depends on the trust between the sending and transmitted agents. Overall social capital can be calculated by the numbers of connections, trust between agents, and information density in a particular time slot. The following formula shows our method to measure social capital measurement in a network with n members:

$TSC\ (t) = \sum SC\ (Rij)$ while $0 \leq i \leq n$ x $(n-1)/2$, $0 \leq j \leq n$ x $(n-1)/2$

TSC: Total social capital
SC: Current social capital
t: At time t
Rij: Relation between agent i and agent j
n: Number of members in the network

The above equation shows that the total social capital can be calculated by all relations' value in the time t. The relation between social capital in time t1 and social capital in time t0 is shown below.

$SC\ (Rij,\ t_1) = SC\ (Rij,\ t_0) + \sum TRij$ x $F\ (Dij)$ x $F\ (Vij)$ while $0 \leq i \leq n$ x $(n-1)/2$, $0 \leq j \leq n$ x $(n-1)/2$

$SC\ (Rij,\ t_0)$ = time + budget + opportunity cost spent to create current the social capital

$F\ (V)$ = time + budget + opportunity cost required to spend to increase social capital in the time slot

$F\ (D)$ = data density share in communication

TR: Trust value

As it is seen in the above equation, trust and information sharing between agent i and agent j and the cost of creating this relation can affect on the total social trust. The method we use to measure social capital is the Investment Method [7] which considers cost that persons spent to create or improve their social capital including time, direct cost, and opportunity cost.

F(D) is the function of knowledge sharing. It depends on various factors including (i) people's marginal propensity to share data which is related to social trust and culture in society; (ii) the level of facilities that help people to share their knowledge, for example virtual environment helps to increase knowledge sharing; (iii) people's interest in the subject, for example some people like to follow political subjects and some people like to follow art news; (iv) knowledge level of sender and receiver agents in which if it is not the same level, it might be boring to both sides to share their knowledge and the efficiency of knowledge sharing will go down.

3 Human Capital Measurement

Human capital is defined in different ways; however, most of its definitions include knowledge as the main source for human capital. Hudson defines human capital as people's experience and education level, their attitudes about life and business, and their genetic inheritance (competency) [8]. Knowledge basically comes from formal education systems that can be calculated easily by the cost of the education, the time that people spent to learn the formal knowledge, and the opportunity cost. Also knowledge comes from casual thinking and it can be named innovation. This kind of knowledge has a great return rate of investment and can be calculated by their added value and their impacts on the business. Trust is very important in human capital as people's attitudes about life and business is related to their trust level for themselves and also for environment i.e. friends, managers, society, etc.

In order to measure human capital, we shall measure knowledge value of education, innovation, and skills. We can calculate knowledge value of education by calculating cost that one spends to gain the knowledge. In this method we suppose that education is a product that we buy and pay for it thus we shall calculate all of the costs involved in the process of gaining education. The main costs in this category are as follows:

Investment – Investment in a formal education system such as cost of education in school, university, and some short term courses or any tuition fee one spends to get formal knowledge.

Time – Time that one spends in the class including studying time and time related to education system.

Opportunity cost – Opportunity cost is related to the cost of the opportunities that one looses due to spending for the education. For example if one continues his/her master and does not work. Thus s/he can not earn money and looses some opportunities.

The second category in human capital is knowledge value of skills. Basically skills are gaioned from experience. In this category the main costs are as following:

Cost of training – this kind of cost is related to the job training, mentoring training and all the costs business firms spend to improve their employee's knowledge.

Cost of experience – practice can improve people's productivity and business firms spend huge amount of money on their employees to increase their experience. This experience is a valuable asset and most of the business firms try to recruit experienced people from their competitors.

Time and opportunity cost – business firms invest in new employee who has just been filled the position to improve knowledge up to the required level. Business firms also lose opportunities in labor market.

The third category in human capital is knowledge value of innovation and it is related to people's competency in innovation and creativity. Although basic knowledge is important in this category, the main important parameter in this category is environment. Dynamic environment can make a suitable environment to enact the people's competency and high level of trust is the important variable in creating this environment. The total value of human capital is the sum of these three categories.

4 Market Capital Measurement

The main part of market capital is knowledge sharing between different components of the business such as suppliers, customers, competitors and others. Bontis defines customer capital as a relational capital which in effect encompasses the knowledge embedded in all the relationships that an organization develops whether it is from customers, from the competitions, from suppliers, from trade associations, or from the government [9,10]. As we have seen in social capital, making and developing relationships is related to trust between agents. Thus, knowledge and trust are the core variables in the market capital. The market capital can be calculated from the knowledge sharing and trust.

In economy, we talk about marginal propensity to buy, sell, or replace and analyze business components. In this paper, we look at viewpoints of running business from knowledge and trust and measure market capital on the basis of our core variables i.e. knowledge and trust.

4.1 Supplier to Company (SPCO)

In the relation between suppliers and company, it relates to marginal propensity to sale (from the business firm view point). The market capital can be calculated as shown below.

> Marginal propensity to sell (t_0) = trust level x (time + money and opportunity cost to make relation) + replacement cost
>
> Marginal propensity to sale (t_1) = Marginal propensity to sale (t_0) + trust in the new relation x (time + money and opportunity cost for the new relation)
>
> Trust in the new relationship depends on the delivery time and also logistic
> $-1 \leq$ trust level ≤ 1

4.2 Supplier to Competitor (SPCM)

When the trust level between the suppliers (or customers) and the business firm is less than the trust between the supplier (or customers) and other competitors, their market can be replaced. The market capital can be calculated as shown below.

> Marginal propensity to replace = trust level for the other companies x (time + money and opportunity cost to make relation) + replacement cost

Trust level for other companies is very important variable in this part and depends on social trust. In liberal market due to the high level of social trust it can be replaced easily.

4.3 Company to Customer (COCS)

This part of business is the main part and businesses can not continue without this relationship. In general situation, people buy goods or use services from a firm when they have enough information as well as they trust the firm. Because of the high cost of the customer replacement for the firms (five times more than royal customers) it is very important that companies focus on this part of business. This capital can be calculated as shown below.

> Market capital for a royal customer = trust level x (time + money + opportunity cost spent to make the current trust - loyal customer) + cost of replacement (to find a new customer + to make loyalty and to improve it up till the trust level)

Increasing the level of the trust depends on the social trust and difference from one culture of location to the others.

4.4 Customer to Customer (CSCS)

This part of the business is going to be the most important part of the business based on the basis of the IT revolution. The new promotion plans such as "word of mouth" is created to improve the embedded market capital in this section. The most important variables in this part are related to knowledge and trust. Knowledge of the customers about a company and knowledge sharing between potential customers in market targets together with trust between them are the most important variables. The market capital in this section can be calculated as shown below.

> $TMC(t_0) = \sum MC(R_{ij}, t_0)$ while $0 \leq i \leq n \times (n-1)/2$, $0 \leq j \leq n \times (n-1)/2$
>
> $TMC(t_1) = TMC(t_0) + \sum TL_{ij} \times F(D_{ij}) \times F(V_{ij})$
>
> TMC: Total market capital
> MC: Current market capital
> R_{ij}: Relation between customer i and customer j
> n: Number of the customers in target markets
> TL_{ij}: Trust level between customer i and customer j
> $F(D_{ij})$: Knowledge sharing density between customer i and customer j
> $F(V_{ij})$: Value of the shared knowledge between customer i and customer j

4.5 Potential Customer to Company (PCCO)

This part of the market capital can be named Brand value. It depends of the trust level of the non-customers and also knowledge of the non customers about the business form. This part of the capital can be created in new markets or new products.

5 Ontology for Intellectual Capital Measurement

Ontology in this paper is an explicit formal specification [11] of the concepts in the domain of intellectual capital of measurement and relationships among concepts. In the ontology, we model human capital, social capital, and market capital and their measurement in detail as stated in the above section. Ontology helps to map intellectual capital measurement and represent a set of relevant concepts with their properties and concept relationships. The ontology helps to come to an agreement on the intellectual capital measurement within an organization.

There are lacks and incompleteness of existing ontologies focusing on intellectual capital measurement in 3 main capital i.e. human capital, social capital, and market capital. The existing ontologies e.g. the enterprise ontology, TOVE, Doblin Core, etc. are too general [12]. They do not take intangibles into account and lack good representation of intangibles [12].

The main role of an ontology for intellectual capital measurement is to represent and share the common understanding of intellectual capital measurement assumptions and the informational structure within an organization.

Figure 1 shows an overview of intellectual capital ontology.

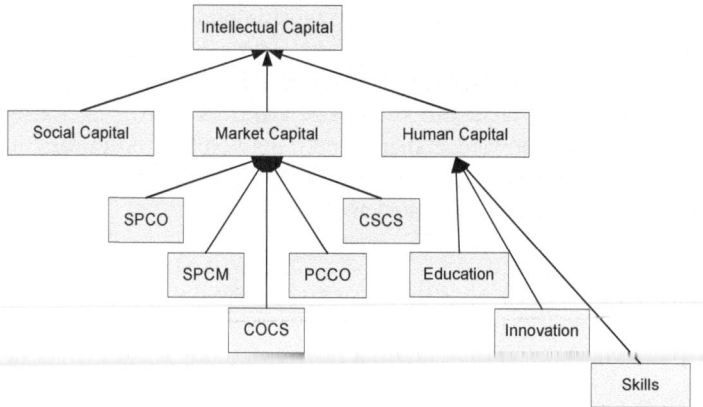

Fig. 1. Overview of Intellectual Capital Ontology

Company, supplier, competitor and customer can be illustrated as agent which relationships can be made between different persons in these kinds of agent. Figure 2 shows agent classification in the ontology.

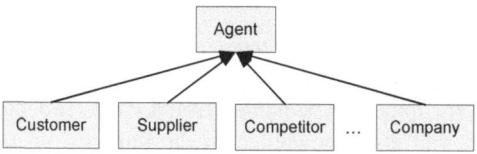

Fig. 2. Classification of agents

In the social capital measurement, the ontology classifies such bodies of knowledge along with a number of properties and constraints i.e. related agent, relating agent, the number of persons in the network, value of a connection, strength of a connection, time slot, kinds of relationships within network, etc. In human capital, the ontology expresses the classification of education, skills, and creation within human resource. Human capital measurement properties and constraints include cost of investment, time, opportunity, etc. In market capital, the ontology classifies into 5 categories i.e. SPCO, SPCM, COCS, CSCS, and PCCO. Generally speaking, an ontology is used to explicitly express how the intellectual capital is valued and thus results are sensible for managers. Additionally, the ontology enables evidence-based formation of knowledge valuation.

6 Conclusion

We have presented intellectual capital constituent and their measurement methods at a conceptual level utilizing a particular type of knowledge based technology, the ontology. Each intellectual capital constituent i.e. social capital, human capital, and market capital has been defined and its measurement has been presented. In our future work we will be focusing its development in each capital.

References

1. Marr, B., Gianni, G., Neely, A.: Intellectual capital-defining key performance indicators for organizational knowledge assets. Journal of Business Process Management 10(5), 551–569 (2004)
2. Fukuyama, F.: Social capital, civil society and development. Third world quarterly 22(1), 7–20 (2001)
3. Fukuyama, F.: Trust. The Social Virtues and the Creation of Prosperity, London (1995)
4. Putnam, R.: Social capital: measurement and consequence (2001)
5. Putnam, R.D.: Bowling alone: America's declining social capital. Journal of Democracy, 65–78 (1995)
6. Deardorff's GlossaryofInternationalEconomics,
   http://www-personal.umich.edu/~alandear/glossary/s.html
   (accessed September 20, 2008)
7. Measuring knowledge assets of a nation: Knowledge systems for development. Department of Economic and Social Affairs. New York, UN (2003)
8. Hudson, W.: Intellectual capital: how to build it, enhance it, use it. John Wiley, New York (1993)
9. Bontis, N.: Managing organizational knowledge by diagnosing intellectual capital: framing and advancing the state of the field. International Journal of technology management 18(5-8), 433–462 (1999)
10. Bontis, N., William, C.C.K., Richardson, S.: Intellectual capital and business performance in Malaysian industries. Journal of intellectual capital 1(1), 85–100 (2000)
11. Gruber, T.R.: A translation approach to portable ontology specification. In: Knowledge Acquisition (1993)
12. Jussupova-Mariethoz, Y., Probst, A.R.: Business concepts ontology for an enterprise performance and competences monitoring. Computers in Industry 58, 118–129 (2007)

e-Health in the Age of Paradox: A Position Paper

William McIver Jr.

National Research Council Canada, Institute for Information Technology
46 Dineen Drive, Fredericton, New Brunswick, E3B 9W4 Canada
Bill.McIver@nrc.ca

Abstract. This position paper examines a critical paradox in e-health: there is a striking gap between critical information services for health care that can be implemented today using existing in information and communication technologies and those services that are actually available. Facets of this paradox are examined in the context of Canadian analyses and policy, advanced research on health care reform, and current technological developments. Hypothetical scenarios are employed as a means of discussing the paradox and, ultimately, of describing potential solutions that are feasible now.

Keywords: e-health, interoperability, patient-centred care.

1   Scenario

Duncan, a computer scientist, was driving an elderly woman, Jessica, and one of her family members, Alia, from Ferney to the Montreux airport in the summer of 2007. During the trip, Jessica, began coughing violently. As Duncan looked over at Jessica in the passenger seat, her head slumped forward and her eyes closed. She, in lay terms, appeared to Duncan to "pass out." He immediately pulled the car over to the side of highway. Shortly after the car had stopped, Jessica opened her eyes and asked what had happened. Jessica appeared not to be in distress of any kind, but the fact that she is a diabetic and over weight made the situation more urgent for Alia and Duncan.

Alia, who is a physician and was acutely aware of Jessica's medical history, began to assess her condition and to make decisions about what she and Duncan should do. Montreux is 177 km from Ferney. The were just over half way to Montreux at that point in a relatively rural part of the Province. It was not yet clear what had happened to Jessica or what her condition was. Should they go forward to one of the two major hospitals in Montreux, which are highly reputable? Or should they return to Ferney, where Alia had far greater knowledge of the health care environment?

They decided to go forward to Montreux, but which hospital? Montreux has the Dr. James Fazy Regional Hospital, which would seem the first choice, and the Montreux City Hospital, another quite capable health care centre. The choice was not so obvious, however. With a sense of urgency, geographic factors weighed heavily in answering this question as they were not at all familiar with Montreux. They did not have a map, but they did have a mobile phone. They soon entered a rest stop and were able to acquire a map of Montreux. They decided that navigating to Fazy Hospital would be relatively easy.

M. Ulieru, P. Palensky, and R. Doursat (Eds.): IT Revolutions 2008, LNICST 11, pp. 240–249, 2009.

The staff at Fazy received Jessica in a professional and courteous manner. She was admitted for tests and observations. While Jessica was made comfortable, a somewhat tedious process ensued with Alia conveying Jessica's insulin regimen to the attending physician. This involved reviewing hand written notes created by Jessica's family and carried by her on her travels. Printouts of prescription information from her pharmacist back in her home province were sought. These were retrieved by another family member and faxed to the hospital after about an hour wait. The two lists had to be reconciled in terms of terminology, quantities, medical history, and other information which differed between the family's informal, though day-to-day, records and those of the pharmacist.

It was early evening when this was finished and they settled in for the night to wait for test results. Jessica was confused and scared, not knowing what had happened to her, much less what was happening in the hospital. Why the wait? What were the physicians doing? Alia, as a physician, was an invaluable resource for Jessica in this context, as she was able to explain to Jessica what was happening. How much more confused would Jessica have been if Alia – a physician -- had not been present? How would that have affected her condition and the task of communicating to the attending physicians about what had happened in the car?

The first attending physician seemed quite open-minded while listening to Duncan and Alia's recounting of the events in the car. The second attending physician who came on duty the next morning seemed to Duncan, a lay person in this context, to convey a much more skeptical, though professional, attitude as they repeated the story. Again, Alia was an invaluable advocate for Jessica in this context. In a way, she, as a proxy for Jessica, was able to achieve a virtual type of patient-centred care where none seemed to exist. As an expert advocate, Alia was able to suggest possibilities to the second attending physician for which diagnostic tests could be sought.

Jessica was released from the hospital two days later. The cause of the event was not conclusively determined. It was felt by Alia that Jessica had experienced a vaso-vagal event, which had caused her to "faint." Because they were not able to cite a definite cause, this episode left Jessica's extended family with a considerable amount of worry for the next month after she finally flew home.

The episode left Duncan, the computer scientist, extremely frustrated by the paradox of information and communication technology (ICT) in this context.

2 Paradoxes

Duncan had actually identified several paradoxes that day. Duncan knew that he and Alia could have easily located a health care facility using widely available and affordable global positioning system (GPS) ICT had they been prepared. Knowing Jessica's general physical condition, Duncan also could have, with a modest amount of money, acquired ICT to monitor and record various of Jessica's vital signs in case the need to provide health care practitioners with that information arose – especially a skeptical attending physician. Duncan knew also that the task of retrieving a current and comprehensive picture of Jessica's medical history and drug regimen could have been achieved using common database management system technology, connecting physicians, pharmacists, hospitals, and other relevant entities. Duncan thought as well that user-centred design and collaborative design approaches -- long in vogue in certain parts of the ICT sector -- could have been leveraged in a straight forward manner in the design of workflow and user interface systems to provide a more patient-centred experience for Jessica and, especially, for others not so fortunate to have a personal physician advocate with them, like Alia. With such a design, hospitals can provide patient-centred experiences, where patients and families are given more information about what is happening, where there is a free exchange of information between patient and physician, where patients can be shown graphical demonstrations that help to explain their condition, or their hypothetical condition until such time as diagnostic tests confirm or deny it. Also, through user-centred design, personal health monitoring systems should now be as easy to use as the telephone for people of Jessica's generation. Duncan knew as well that common social networking, video and other common Web technologies could have been employed to leave Jessica with a summary of her experience, the hospital's findings, and information to help her care for herself in the days that followed as she returned home.

The gaps between what Duncan knew was possible with current ICT and what actually existed within the health care situation Jessica found herself that day constitute a paradox in healthcare. Charles Handy, a well known philosopher of management, has posited that we are in an age of paradoxes, wherein traditional practices confront new situations brought about by rapid advances in some aspect of society [1]. Two of the nine paradoxes Handy has defined are most closely related to the paradox identified by Duncan. These are a paradox of intelligence, where the means of production is no longer dependent as much on physical labour as it is on the knowledge that labourers hold; and a paradox of work, where a few people are over worked while many others are unemployed. Handy's paradoxes speak to broad societal conditions; Duncan's paradox merely defines healthcare-specific variations on Handy's paradoxes. We will call Duncan's paradox the paradox of eHealth.

3 Building Blocks

The paradox of eHealth is not new and has been the attention of many entities for a long time. The "binding sites" for the unravelling of this paradox can be seen, in particular, in the first three of the "four essential building blocks" called for by the *Romanow Commission on the Future of Healthcare in Canada* [2], the last major review of Canada's healthcare system. Romanow's four pillars are "continuity of care, early detection and action, better information on needs and outcomes, and new and stronger incentives for health care providers to participate in primary health care approaches." They are also in the foundation, the *Canada Health Act* [3], which establishes the criteria for Canada's provinces and territories to provide medicare as public administration, comprehensiveness, universality, portability and accessibility.

It is important to point out that despite Duncan's confusion, Canada enjoys a good health care system. As has been written by Michael Rachlis, M.D. [4], Canada's most widely-recognized health policy analyst, Canadians "still believe in the values of medicare." Nonetheless, they want a number of things fixed or improved. The Romanow Commission was established in 2001 by the Prime Minister of Canada with the goal of seeking policy recommendations for ensuring "the sustainability of a universally accessible, publicly funded health system, that offers quality services to Canadians and strikes an appropriate balance between investments in prevention and health maintenance and those directed to care and treatment." Among its many recommendations, the Commission called for the following:

- improving access to health care in rural and remote areas;
- reducing wait times for diagnostic services;
- facilitating the transfer of patients to home care when appropriate;
- the establishment of a system for managing personal electronic health records across Canada; and
- improving the coordination of "staffing of physicians, nurses, and other staff in areas where they are needed."

4   What Exists?

The building blocks for comprehensive solutions to fulfill the Romanow Commission's recommendations already exist. Building blocks that take advantage of current generation mobile wireless data communications and service-oriented software architecture technology, including Web services, have been emerging over the past few years. This convergence is being called telemedicine 2.0 [5].

Telemedicine 2.0 is being defined as the "convergence of clinical and personal wireless solutions," where traditional clinical telemedicine enables delivery of care to fixed locations [5]. In this context, telemedicine 2.0 integrates traditional fixed location devices with mobile systems within some comprehensive data management scheme. The mobile phone is seen as the focal point of this convergence. These systems are characterized into the categories of Monitoring, Software and Services, and Tracking. Some telemedicine 2.0 solutions are focused on single disease management, while some are emerging that have multiple disease management capabilities. CardioNet [6] is one example of a focused solution. It provides mobile outpatient cardiac telemetry services. Toumaz Technology [7] of Abingdon in the U.K. is developing a wireless medical monitoring system, called the Sensium, within a "band-aid" form factor that they call a "digital plaster." Multiple sensors can be linked together into a wireless body area network using their technology.

Many of these services employ Mobile Virtual Network Operators (MVNO). They do not own spectrum or infrastructure. They lease these services from mobile telephony and Internet network providers and focus on the provision of their specialized services.

Packet radio protocols exist for the major mobile telephony protocols. These include General Packet Radio Service (GPRS) or Enhanced Data rates for GSM Evolution (EDGE) within the set of Global System for Mobile communications (GSM) 2G mobile telephony standards; High Speed Packet Access (HSPA) within the Universal Mobile Telecommunications System (UMTS) set of 3G mobile telephony standards; and Evolution-Data Optimized (EVDO) within the CDMA2000 set of mobile telephony standards. Efforts are being made to unify GSM and non-GSM protocols as well. Though the choice of packet radio protocols can be encapsulated somewhat through a focus on the Internet Protocol (IP) level when architecting software applications.

5 ICT's Further Potential

The specific characterization of the paradox of eHealth is that current ICT offer the constituent elements of solutions to the problems within the health care system, but they are, for various reasons, slow to be leveraged in a comprehensive way. ICT should not be seen as solutions, only parts of solutions within a responsible approach to systems analysis and design. At an abstract level, and without oversimplification, solutions to the eHealth Paradox must address, in part, the following issues:

- coordination of complex and extensive work flows,
- distributed and secure information sharing,
- management of distributed staff;
- location and tracking of people and objects; and
- providing customized client-centred service.

Workflow management, distributed database management systems, workforce management, location-based services and object tracking are all mature technologies. Interestingly, they are applied successfully by a number of organizations in other domains besides health care that have requirements similar to those listed above and which have addressed these requirements in comprehensive ways. Some have even achieved *Six Sigma* performance in doing so, or so they have claimed. Interestingly, these include organizations such as airlines, postal systems, and couriers.

Healthcare cannot be equated directly to the means by which these types of organizations provide their services. Software systems cannot by themselves "respect the individuality, values, ethnicity, social endowments, and information needs of each patient," as Rachlis frames the rules for patient-centred care [4]. Nonetheless, healthcare still overlaps in significant ways with the requirements identified above in terms of the types of data communications and operations that are necessary to manage information about a patient.

Rachlis also points out that patient-centred care must include "continuous relationships, care customized to the patient's needs, care controlled by patients, [giving] patients unrestricted access to their records, sharing of knowledge freely between a patient and their different providers using the latest ICT" [4]. These criteria do not differ fundamentally from the functionality provided by the best of the current client relationship management (CRM) systems. Integrated properly into a health system, such ICT could provide elements of solutions that satisfy Rachlis's criteria for patient-centred care.

While key ICT building blocks and mature organizational strategies have been developed, major barriers to creating comprehensive eHealth systems remain. The major barriers are organizational inertia, and ICT and data interoperability.

6 Jumping Organizational Barriers

Canada Health Infoway, a federally-funded, independent, not-for-profit organization whose mandate is to "accelerate the use of electronic health information systems and electronic health records (EHRs) across the country" has been working since 2001 on some of the requirements described in the previous section [8]. Canada has thirteen jurisdictions in which the Health Infoway plan is being implemented. Three jurisdictions were found to be on track to implement an EHR by 2010. Two jurisdictions are predicted to implement a full EHR model during 2010. Four jurisdictions are predicted to have only core EHR functionality in place by 2010. Four jurisdictions will require significant time beyond 2010 to implement core functionality for an EHR.

It was found in a 2007 review of Infoway's progress that the barriers to its vision have not been related to technological issues so much as they have been policy-related. The report found that Infoway made "inconsistent and insufficient commitments" and that they demonstrated an "inability to fully illustrate the impact" of the changes that are required [8].

Interestingly, Rachlis's advice for how to implement change within the justifiably skeptical and conservative domain of health care almost mirrors what are thought to be best practices within the now hot field of design thinking. Design thinking processes are applied to the creation of innovative technologies and services in such a way that the user's (e.g. patient's) concerns and perspectives drive the design activities and ultimately guide the selection of solutions. IDEO CEO Timothy Brown demonstrated this in a 2006 lecture at MIT [9]. He showed how design thinkers, including psychologists, anthropologists, engineers, and people from a wide variety of other disciplines can produce successful designs which are appropriately situated within the given social and organizational environment. This has included putting design staff on hospital gurneys with video cameras so that they can capture authentic patient experiences. This latter practice lead to the insight that patients are often given very little information about what is being done to care for them from that vantage point. Brown suggested the placement of a status monitor on the ceiling above each patient as the beginning of a solution.

Rachlis's advice is specific to health care organizations in this context, however, recognizing that they are often reluctant to try new ideas. Changes, in Rachlis's approach, should be made in incremental ways as follows [4]:

- it must be easy for health care providers to "dip their toes into the water of innovation";
- providers need the ability to see systems in operation first;
- systems must also fit within the social context in which they are to be employed; and
- changes should be tested using methods such as the Institute for Healthcare Improvement's Plan-Do-Study-Act cycle.

Most interestingly, Rachlis also calls for maximizing the "talents" of patients and providers in facilitating improvements in health care [4]. This is a key design thinking approach.

Helping health care providers "dip their toes into the water of innovation" is the role of independent organizations such as the National Research Council of Canada (NRC). Commercial forces, while necessary for innovation and providing the implementation that are ultimately to be employed, can erect barriers to standardization and interoperability, as von Hippel [10][11] has shown, as a side-effect of their maintenance of proprietary information about their products and services.

Handy's paradoxes of intelligence and work both speak to the need to improve the coordination of staffing of health care teams. In many cases, when patients are in rural and remote areas lacking access to both family doctors and specialists, this has and will continue to require tele-medicine approaches. However, current ICT makes possible far more than what has been implemented in the past. This includes the recording and manipulation of video by patients themselves using easy to obtain Web-based technologies, known as peer-generated video (PGV). PGV looks to be a promising way for patients to communicate with remote care givers. Tele-work and other collaborative technologies also make it possible for health care organizations to balance work loads among existing professionals as well as providing ways to educate more practitioners to increase access to care for all. For example, distance education employing the latest in high definition video and collaborative virtual learning environments make possible the training of physicians, nurses, and diagnostic specialists independent of their residence. This spanning of geographic barriers may be one key to unlocking as of yet unrealized health care labour potential within Canada.

Finally, in the context of organizational barriers, technical capabilities cannot be viewed apart from the political-administrative models of healthcare. Much of the telemedicine 2.0 development seems to be driven by U.S. healthcare companies. They assume a privatized model of healthcare, whereby a health care organization buys into their particular solution. Solutions for a Canada Health Infoway may not be well-served by such an assumption. Universality requirements and norms as well as economies of scale would probably be best served by standardization directed by public processes and not by industry, as this has historically tended to steer technologies away from complete interoperability.

7 Jumping Interoperability Barriers

The work of Canada Health Infoway has gone a long way toward reducing interoperability barriers at the data level and the health management systems level. Much of their effort has involved standardization around the Health Level 7 (HL7) [12] body of standards. HL7 is perhaps the most mature component of all of the issues raised in this paper. It has been in development since 1987. It has evolved to HL7 version 3 (v3). Infoway has adopted HL7 v3 [13].

HL7 v3 is an XML-based standard that allows for both human and machine readable exchange of information [12]. This capability provides a key component for the alternate scenario to Duncan's story, which is described in the next section. HL7 v3 defines both a Clinical Document Architecture (CDA), which evolved out of the earlier Patient Record Architecture, and a Messaging Standard. Clinical documents may contain historical data, medical diagnostic data, narrative information, and should be signed or "attested" using digital signatures. The messaging standard emphasizes machine processability, though they can include human readable information. Both models may be employed together. Recommendations are made by HL7 as to which model to employ in a given use case.

The most serious interoperability barriers seem to remain between devices and software services. Interoperability may exist within silos defined by individual manufacturers or services, but real solutions require broad interoperability across data, device, and health management system layers. The U.S. Food and Drug Administration reported in 2005 that there were 25 blood glucose monitors on the market, each with their own data protocols [14]. It is often not clear whether these devices and, at a higher level, telemedicine 2.0 systems are based on open standards. Thus, individuals and institutions that adopt one set of devices may be locked into a particular MVNO or data communication protocol and, thus, lack the possibility of interoperation with other devices they may require.

Some device manufacturers and MVNOs are responding through the creation of gateways that support an extensible list of devices. The company 4HomeMedia offers an open standards "health hub." Among their digital home products, they provide OEM-oriented solutions for enabling consumer electronic (CE) devices in the home. According to Engaget.com this includes a health monitoring solution [15]. The company LifeComm was announced as a partnership between Qualcomm, the dominant mobile telephony chipset and software provider for the Code Division Multiple Access (CDMA) protocol. LifeComm is working with handset manufacturers to develop mobile telephony-based MVNO hardware and software for a class of monitoring devices [16]. Another multiple disease management system provider is *iMetrikus* [17]. They advertise a "low-cost" gateway to channel personal health monitoring data into clinical systems. They claim that their *MediCompass Connect* system can interface with over 45 different devices. This can be done via IP-based data communications on a personal computing device or via a telephone connection.

At a data level, Google Health [18] allows individuals with Google accounts to import, organize and maintain their own medical records. Facets of one's PHR that can be maintained or automatically generated by Google include the following: medical profile, including conditions, medications, allergies, procedures, test results, and immunizations; notices sent from health providers; and drug interaction information generated by Google based on one's medication profile. Google Health offers the ability to synchronize Google PHRs with other electronic medical record (EMR) sites, including *Epocrates Patient Snapshot, iHealth, Lifestar, MyMedicalRecords.com*, and *NoMoreClipboard*. The capabilities of Google's search technology and its high profile as a central access point to the Web for many people has engendered security and privacy concerns over Google Health.

Interoperability in the end may be best served by having health organizations guide the definition of application programming interfaces and hardware interfaces through public and industry-independent processes which allow corporate input, but which prioritize the best interests of public health systems above other interests.

8 An Alternate Scenario

Duncan imagines the following scenario, all enabled by current ICT. Because Jessica is at risk, her healthcare team has fitted her with a watch containing a heart rate monitor that stores its data on an internal flash drive. Jessica is also carrying a bluetooth-enabled blood glucose meter. All of her monitoring devices are interoperable at a physical layer through common interface standards such as Bluetooth and USB to permit easy transmission of data to her mobile phone and other common computing platforms. This level of interoperability allows the data to be easily delivered to health care providers in situations such as her fainting event in the car. All data formats and exchanges are standardized on either HL7 clinical documents or HL7 messages, as appropriate, providing for interoperability at the data level.

Besides being Bluetooth-enabled, her mobile phone is also capable of using Session Initiation Protocol (SIP), which allows it to establish sessions with the regional health authority's SIP-enabled E911 service. Jessica's mobile phone uses SIP to notify the regional health authority's patient care workflow system of its data communication capabilities, its location, and of all of the types of data about Jessica that it has available. SIP is then employed to negotiate the delivery of these data to the authority. The voice channel on the phone would be employed concurrently by Alia to communicate with an emergency room physician while Duncan continued driving.

The regional health authority would receive Jessica's medicare number within the encrypted SIP message, create a new case file, and place it into its dynamic patient care workflow engine. The workflow engine is capable of managing health care algorithms selected by the health care team and can deliver associated information to appropriate "nodes" within the regional health care authority as Jessica receives care. This could include retrieving and routing her patient health record to the emergency room desk so that an attending physician can review it before she arrives. If necessary, the physician could use the system to call Jessica's phone to consult with her while she is still in route.

Using a decision support system in conjunction with the workflow system, the regional health authority is able to identify available capacity within an appropriate hospital. The authority then arranges the delivery of a map to the cell phone showing detailed directions to the chosen hospital.

Before Jessica arrives at the emergency room, she has undergone a preliminary triage based on both the data received from her mobile phone and communications with Alia in the car. A formal triage is completed on her arrival, but has been made more efficient by information that was made available about her ahead of time.

Schema mapping facilities provided by the distributed database management system are employed to reconcile information provided by the family, such as adjustments to Jessica's insulin regimen made by the family since she last visited her physician. The view mechanisms of the system would be employed by physicians to examine Jessica's patient record in any number of ways. In addition, analytic services provided by the system can be performed over any of the data sets in Jessica's record, including the recent data sent from her mobile phone. These might include new analytic techniques that have been validated by the latest medical science results.

Inside of Fazy hospital, Jessica is given a small, flat electronic paper display by her bed that shows her information about what her health care team is doing for her. The information can be delivered by voice, if appropriate, and in different languages. A virtual learning environment is also available through the electronic paper for Jessica and her family members to retrieve explanations of her condition and of her diagnostic test results. This learning material is structured so that it can be delivered at various levels of complexity. Alia is able to choose an article and graphical demonstration of vaso-vago responses at a level appropriate to Jessica's level of understanding.

Before Jessica is released from the hospital, her Personal Health Portal is synchronized with HL7 CDA-formatted data about the following:

- information about her hospital visit,
- what the findings of the physicians were,
- updates to her patient health record, and
- video-based instructions by her health care team about how she should care for herself at home.

When she arrives back in her home province, her family members help her to sign on to her personal health portal. They all feel much more comfortable about caring for Jessica than they would without having this information.

References

1. Handy, C.: The Age of Paradox. Harvard Business School Press, Boston (1994)
2. Romanow, R.J.: Building on Values: The Future of Health Care in Canada, Commission on the Future of Health Care in Canada, iii, xxviii, 110 (November 2002)
3. Government of Canada, Canada Health Act (R.S., 1985, c. C-6) Sec. 7 (a) - (e) (1985), http://laws.justice.gc.ca/en/C-6/
4. Rachlis, M.: Prescription for Excellence: How Innovation is Saving Canada's Health Care System, Harper Collins Canada, 22, 62-63, 334 (2005), http://www.michaelrachlis.com/pubs/Prescription%20for%20Excellence%20pb%200412.pdf
5. Triple Tree, TELEMEDICINE 2.0: Connecting Medical Devices, Patients and Providers to Improve Health: A TripleTree Industry Analysis, 3 (2007), http://www.triple-tree.com
6. CardioNet, http://www.cardionet.com/
7. Toumaz Technology, http://www.toumaz.com/products/sensium.htm
8. Canada Health Infoway, Vision 2015 Advancing Canada's Next Generation of Healthcare, 4 (2007), http://www.infoway-inforoute.ca
9. Brown, T.: Innovation Through Design Thinking, MIT World, video (2006), http://mitworld.mit.edu/video/357/
10. von Hippel, E.: Sources of Innovation. The MIT Press, Cambridge (1988)
11. von Hippel, E.: Democratizing Innovation. The MIT Press, Cambridge (2005)
12. HL7, CDA Release 2.0 DRAFT (April 08, 2002), http://www.hl7.org
13. Canada Health Infoway, Strategic Direction for Standards, http://www.infoway-inforoute.ca (n.d.)
14. Food, U.S.: Drug Administration. Department of Health and Human Services, Glucose Meters & Diabetes Management (June 14, 2005), http://www.fda.gov/diabetes/glucose.html
15. Murph, D.: 4HomeMedia's HealthPoint 1500 provides remote health monitoring, Engaget.com (November 13, 2007), http://www.engadget.com/2007/11/13/4homemedias-healthpoint-1500-provides-remote-health-monitoring/
16. Smith, B.: Health Care MVNO Planned, WirelessWeek (May 15, 2007), http://www.wirelessweek.com/article.aspx?id=147854
17. iMetrikus, http://www.imetrikus.com/solutions.html
18. Google Health, https://www.google.com/health

Bio-Intelligence: A Research Program Facilitating the Development of New Paradigms for Tomorrow's Patient Care

Sieu Phan, Fazel Famili, Ziying Liu, and Lourdes Peña-Castillo

Institute for Information Technology,
National Research Council Canada, Ottawa, Ontario, K1A 0R6, Canada
{sieu.phan,fazel.famili,ziying.liu,lourdes.pena-castillo}@nrc-cnrc.gc.ca

Abstract. The advancement of omics technologies in concert with the enabling information technology development has accelerated biological research to a new realm in a blazing speed and sophistication. The limited single gene assay to the high throughput microarray assay and the laborious manual count of base-pairs to the robotic assisted machinery in genome sequencing are two examples to name. Yet even more sophisticated, the recent development in literature mining and artificial intelligence has allowed researchers to construct complex gene networks unraveling many formidable biological puzzles. To harness these emerging technologies to their full potential to medical applications, the Bio-intelligence program at the Institute for Information Technology, National Research Council Canada, aims to develop and exploit artificial intelligence and bioinformatics technologies to facilitate the development of intelligent decision support tools and systems to improve patient care - for early detection, accurate diagnosis/prognosis of disease, and better personalized therapeutic management.

Keywords: Information technology, patient care

1 Introduction

The emerging omics technologies have propelled the advancement of our understanding of disease development and progression mechanisms at the molecular basis. This new understanding could potentially lead to the development of new therapeutic targets, paradigms in disease treatment, and drug design. The Bio-intelligence program (Figure 1) at the Institute for Information Technology (IIT), National Research Council Canada (NRC) aims to develop novel IT methodologies to harness these emerging technologies to their full potential to medical applications. The Bio-intelligence program is a collaborative effort among NRC-IIT, NRC-BRI (Biotechnology Research Institute), NRC-IBS (Institute for Biological Sciences), the Ottawa Hospital, and Universidad Politécnica de Madrid. The research program consists of two principal components: knowledge discovery and knowledge integration.

M. Ulieru, P. Palensky, and R. Doursat (Eds.): IT Revolutions 2008, LNICST 11, pp. 250–253, 2009.

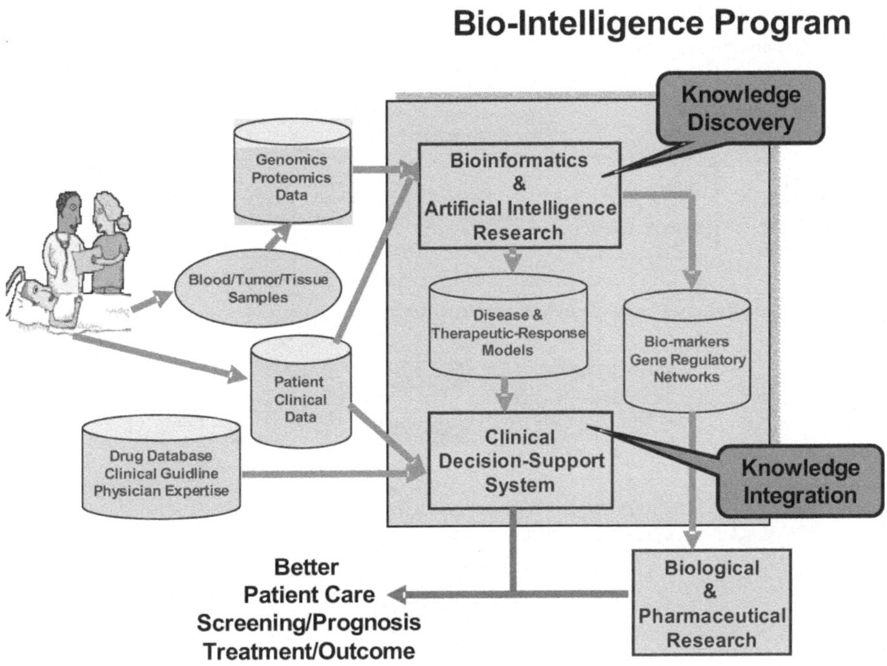

Bio-Intelligence Program

Fig. 1. Bio-intelligence program overview

2   Knowledge Discovery and Knowledge Integration

In the next two sections we discuss the two components of the Bio-intelligence program.

2.1   Knowledge Discovery

Cancer is one of the deadliest diseases. As reported by the Canadian Public Health Agency: "Two in five Canadians face a cancer diagnosis in their lifetime. In 2007 alone, an estimated 159,000 new cases of cancer and 72,700 deaths from cancer will occur in Canada. The burden of cancer in Canada is enormous, affecting the economic and social well-being of individual Canadians, their families and the country". To scale our research to a manageable level according to our resources, our initial piloting effort is focused on cancer treatments.

Cancer is a collection of many diseases in which cells of an organ or tissue in our body become abnormal, growing and multiplying out of control. Normal cells have a life cycle and they reproduce themselves throughout the body in an orderly and controlled manner. Normal growth continues throughout life to replace worn out tissue, to heal wounds, and to maintain healthy organs. When cells grow out of control, they usually form a mass, called a tumour, which results in one form or the other, in cancer. There is a chain of complex processes maintaining a cell's growth and life. In the process of making protein, the DNA

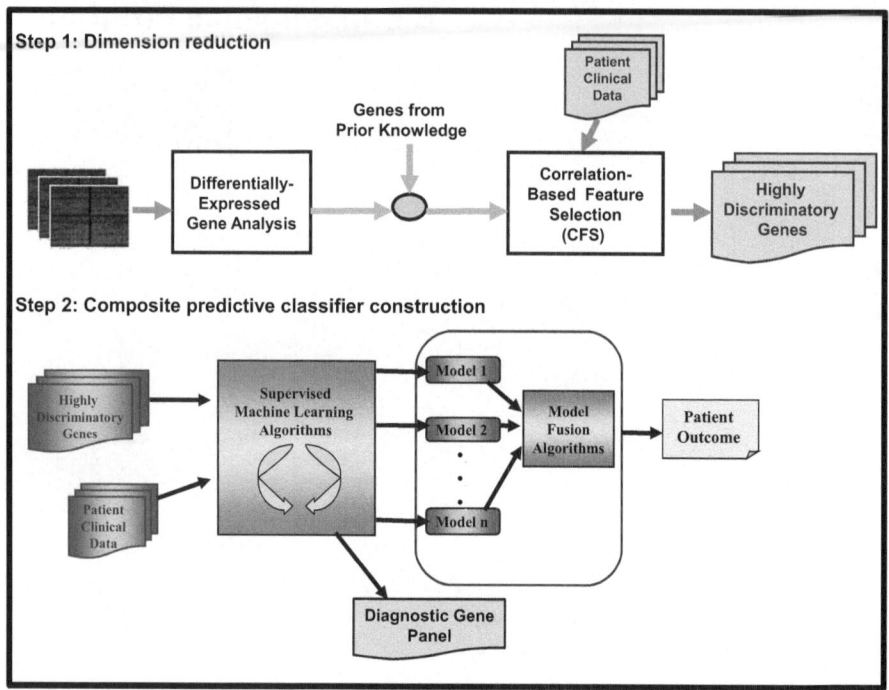

Fig. 2. Workflow for the identification of diagnostic gene panel from transcriptomics data and its companion decision support software module

in the cell nucleus is first transcribed to mRNA; mRNA is then translated into proteins. Some proteins will undergo further post-translational modifications. If the DNA gets mutated or something goes wrong in any of these processes, the cell could develop abnormal behaviour and trigger the development of diseases such as cancer.

Our goal is to develop IT methodologies coupling with the emerging omics technologies to identify discernible DNA/RNA/protein/metabolite variations that could be used to facilitate the fabrication of clinical test kits for early detection, accurate diagnosis/prognosis of cancer, and for the screening of patients' response to cancer therapy. A sample workflow for the identification of a dignostic gene panel from transcriptomics data and its companion decision support software is shown in Figure 2. Our expertise in artificial intelligence, knowledge discovery from data [1,2] and our previous successes in developing intelligent diagnostic and maintenance systems [3] position us very well to take a new approach (than the limited statistical approach) in tackling this life science challenge.

2.2 Knowledge Integration

The second objective of the project is knowledge integration. The plan is to exploit IT to develop a comprehensive clinical decision support system that integrates several sources of knowledge and data. In addition to the usual capabilities such as disease screening, diagnosis, and prognosis, the system will be designed to help reducing medical errors and to facilitate in providing better and more personalized therapeutic management. For example, with the availability of drug databases and patient electronic health record, drug contradiction or adverse side effects could be avoided. With the availability of omics therapeutic response models and personal genetic profiles, more effective personalized treatment plan can be prescribed. Another example is, in the emergency department, where the clinical guideline module could help in speeding up the triage process in the less-frequently encountered situations. In addition, we would like to incorporate an intelligent look-up-and-reasoning module to support evidence-based medicine [4].

3 Conclusion

The emerging omics technologies have advanced biological research to a new frontier - from the limited view of the traditional approach of examining specific genes, proteins, or metabolites in isolation to the broader interaction view of the systems biology approach. This provides a more comprehensive means to study and understand disease development and progression and consequently promises new effective mechanisms to treat diseases and develop new drugs. Our research aims to develop the bridging IT methodologies to facilitate the translation of omics research to tomorrow's much more effective and efficient new paradigms in disease treatment and drug development.

References

1. Phan, S., Famili, F., Tang, Z., Pan, Y., Liu, Z., Ouyang, J., Lenferink, A., McCourt-O'Connor, M.: A Novel Pattern Based Clustering Methodology for Time-Series Microarray Data. Int. Journal of Computer Mathematics 84(5), 585–597 (2007)
2. Famili, F., Phan, S., Liu, Z., Pan, Y., Djebbari, A., Lenferink, A., O'Connor, M.: Discovering Informative Genes from Gene Expression Data: A Multi-Strategy Approach. In: The 18th European Conference on Machine Learning, Warsaw, Poland (2007)
3. Halasz, M., Dubé, F., Orchard, R., Ferland, R.: The Integrated Diagnostic System (IDS): Remote Monitoring and Decision Support for Commercial Aircraft - Putting Theory into Practice. In: Proceedings of AAAI 1999 Spring Symposium on AI in Equipment Maintenance, Palo Alto, California, USA (1999)
4. Centre for Health Evidence, http://www.cche.net

An Integrative Bioinformatics Approach for Knowledge Discovery

Lourdes Peña-Castillo, Sieu Phan, and Fazel Famili

Institute for Information Technology,
National Research Council Canada, Ottawa, Ontario, K1A 0R6, Canada
{lourdes.pena-castillo,sieu.phan,fazel.famili}@nrc-cnrc.gc.ca

Abstract. The vast amount of data being generated by large scale omics projects and the computational approaches developed to deal with this data have the potential to accelerate the advancement of our understanding of the molecular basis of genetic diseases. This better understanding may have profound clinical implications and transform the medical practice; for instance, therapeutic management could be prescribed based on the patient's genetic profile instead of being based on aggregate data. Current efforts have established the feasibility and utility of integrating and analysing heterogeneous genomic data to identify molecular associations to pathogenesis. However, since these initiatives are data-centric, they either restrict the research community to specific data sets or to a certain application domain, or force researchers to develop their own analysis tools. To fully exploit the potential of omics technologies, robust computational approaches need to be developed and made available to the community. This research addresses such challenge and proposes an integrative approach to facilitate knowledge discovery from diverse datasets and contribute to the advancement of genomic medicine.

Keywords: Bioinformatics, knowledge discovery, genetic diseases.

1 Introduction

Large-scale omics experiments are continuously being performed providing a constant source of a huge amount of data. The sheer accumulation of this data have required the development of computational approaches to deal with the collection, storage, integration, analysis, visualization and dissemination of these data sets. The goal of these computational approaches is to enable researchers to advance their understanding of the physiological purpose or molecular role of human proteins, with a special emphasis in better comprehending the molecular basis of genetic disorders. Integration of heterogeneous data sources is of paramount importance to obtain a global view of the molecular associations to multifactorial diseases such as Alzheimer's disease, diabetes, and cancer. These molecular associations could lead to new therapeutic targets, new diagnostic tests, new drug design, and ultimately the transformation of the medical practice.

Current efforts have shown the feasibility and utility of integrating and analysing heterogeneous omics data to identify molecular associations to patho-genesis [1,2]

M. Ulieru, P. Palensky, and R. Doursat (Eds.): IT Revolutions 2008, LNICST 11, pp. 254–257, 2009.

and to predict gene function [3]. Most of these initiatives are data-centric in the sense that their focus is to provide access to high-quality data and to allow researchers to explore this data. These initiatives are highly valuable resources and provide a crucial service to the research community; however, since their focus is to disseminate data and to facilitate the analysis of this data, they limit researchers to a specific application domain (e.g., cancer instead of other genetic disorders) and to specific data sources. To fully exploit the potential of omics technologies, robust computational approaches need to be developed and made available to the community. The recently NCI-launched initiative, caBIG [4] has among its goals the development and dissemination of a compendium of freely available software applications. However, to the best of our knowledge, in caBIG's current tool collection an integrative approach offering a unified framework from data integration to knowledge discovery is missing.

Here, we propose the development of an integrative computational approach to facilitate knowledge discovery from heterogeneous omics data sets that provides a unified framework from data integration to the application of multiple machine learning algorithms to derive classification or prediction models. In addition, this approach is suitable for a variety of application domains (e.g., various diseases).

2   An Integrative Approach to Facilitate Discovery of Biological Knowledge

Our pipeline provides an unified framework to support the following tasks:

1. creation of a data collection which may include data from high throughput studies and available annotation data,
2. query-based dynamic data integration,
3. selection of candidates (genes, functions, pathways, protein complexes, etc.) of interest, and
4. model construction.

Our integrative approach for biological knowledge discovery distinguishes itself from other approaches in:

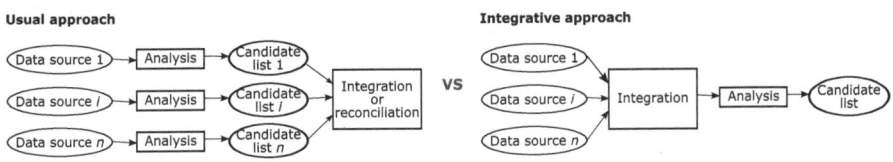

Fig. 1. Comparison of the usual integration approach vs our integrative approach

- performing integration of the heterogeneous data sources before analysis as shown in Figure 1, and
- allowing various abstraction levels during analysis as depicted in Figure 2.

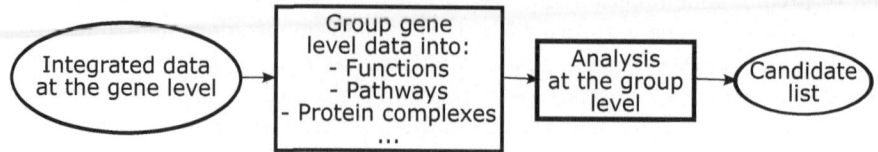

Fig. 2. Analysis at various abstraction levels

By performing integration of diverse evidence available before analysis, we increase the sensitivity of our approach, improve the confidence in the list of candidates since the list is based on several sources of evidence, and are able to observe biological relationships; for instance, we observe a mutation affecting a certain gene and the differential expression of that gene. By analyzing and selecting candidates at various abstraction levels, our list of candidates directly consists of the groups (genes, functions, pathways, etc) of interest.

The main advantages of our approach are a sound methodology for data integration and dealing with missing data; various abstraction levels during analysis (e.g., gene-based or pathway-based); three-state classification models (models include an inconclusive state for borderline cases), and strategies for evaluation of multiple models.

3   Conclusion

To fully exploit the potential of omics technologies, robust computational approaches need to be developed and made available to the community. Several current efforts are focusing on the development of these computational approaches. We propose an integrative computational approach supporting the complete data analysis workflow from data integration to knowledge discovery. Our approach will allow researches to use diverse data sets of their choice and be suitable for various application domains. The end goal of our project is to facilitate the identification of molecular associations to pathogenesis. These associations may lead to the identification of new targets for better diagnosis tests, drugs and, ultimately, to personalized medical care.

References

1. Rhodes, D., Kalyana-Sundaram, S., Mahavisno, V., Varambally, R., Yu, J., Briggs, B., Barrette, T., Anstet, M., Kincead-Beal, C., Kulkarni, P., et al.: Oncomine 3.0: Genes, Pathways, and Networks in a Collection of 18,000 Cancer Gene Expression Profiles. Neoplasia 9(2), 166 (2007)
2. The Cancer Genome Atlas Research Network: Comprehensive genomic characterization defines human glioblastoma genes and core pathways. Nature 455(7216), 1061–1068 (2008)

3. Peña-Castillo, L., Tasan, M., Myers, C., Lee, H., Joshi, T., Zhang, C., Guan, Y., Leone, M., Pagnani, A., Kim, W., et al.: A critical assessment of Mus musculus gene function prediction using integrated genomic evidence. Genome Biology 9(suppl. 1:S2) (2008)
4. Wolfson, W.: caBIG: Seeking Cancer Cures by Bits and Bytes. Chemistry & Biology 15(6), 521–522 (2008)

Author Index